VERMIN CO
WITH TH
AIR RIFLE

JIM TYLER

PETER ANDREW PUBLISHING COMPANY

First edition 1988

Published by
PETER ANDREW PUBLISHING COMPANY

ISBN 0 946796 25 4

Typeset in Great Britain by
Steven Graphics,
Droitwich, Worcestershire.

Dedicated
To the Sporting Air Rifle team.

PREFACE

There are estimated to be in the region of four million air rifles in circulation within the UK; most are purchased nowadays for the control of vermin.

The book represents the first serious treatment of this subject. The author is widely recognised to be the leading authority on the subject, having lectured *per pro* shooting organisations and having been the editor of Sporting Air Rifle magazine.

The book will be of interest to the farming community, to game shooters, to pest control companies and to all airgun owners.

J. Tyler
June, 1988

CONTENTS

Chapter

Chapter 1

Introduction

In the pursuit of farming, forestry and even in the planning and building of housing and factories, man has created environments which are ideally suited to certain species of bird and mammal which flourish to such an extent that they are deemed by the Ministry of Agriculture, Fisheries and Food as 'Vermin'. These species flourish at the expense of other creatures, of food production and in some cases of damage to buildings or the spreading of disease. They are not called vermin because we don't like the look of them: they have to both earn the title in one of the ways described above and they have to exist in such numbers that they do constitute a real problem. No species in danger of extinction, no species in need of protection is ever awarded the title vermin.

Despite the status of such species, they all enjoy certain legal protection insofar as they may only be controlled using approved methods. For instance, it is legal to cull woodpigeon using a gun but highly illegal to attempt to lime, poison, trap or angle for (as in fishing) the birds. Even the much-hated and feared brown rat may only be taken using approved methods.

The air rifle is but one of a range of tools which may be employed in the control of vermin; its effectiveness depends upon the individual circumstances. In the control of a farmyard rat population, for example, the air rifle will certainly kill the rat but it is unlikely that a shooting regime could ever eradicate the problem. In this case, the air rifle should be viewed as a useful supplement to poisons - especially as anti-coagulants disorientate rats so that they appear and may be shot in daylight. In the same way, the air rifle is no solution to a destructive flock of woodpigeon, though on the other hand the air rifle can handle a garden or perhaps an orchard rabbit infestation, feral pigeons on industrial premises or many other specific vermin problems.

For the housholder with a large garden, for the smallholder and for the pest control operative the air rifle is an invaluable tool. For the farmer, forest ranger and the game keeper the air rifle is well worth carrying for opportunist culling of unwanted species.

The greatest strength of the air rifle is often mistakenly perceived as its biggest drawback; namely, the low power. To put the power of the air rifle into perspective, a normal twelve bore game cartridge gives roughly 130 times the power, a sub-calibre .22" rimfire rifle gives anything between six and thirteen times the power. This very low power permits the safe use of the air rifle in many places where other guns would be ruled out as far too dangerous. The air rifle is also very quiet in operation and may be used in a garden without annoying neighbours or on industrial premises without alarming anyone. In many applications, the quiet discharge of the air rifle will permit many shots to be taken whereas a single shotgun blast will clear the whole area!

The low power of the air rifle will only permit kills if the pellet can be quite precisely placed into a usually small target area - near misses will only wound. This firstly demands a high standard of marksmanship, and secondly a thorough knowledge of the quarry in order that close proximity and hence short ranges may be gained.

Whilst on the subject of air rifle power, for years, people have played with standard air rifles in the hope of raising their power - usually with disasterous results. Over-large springs usually do little but increase recoil, reduce accuracy and increase wear. The problem with more powerful air rifles (for which a Firearms Certificate is required) is that the pellet is the limiting factor in the effectiveness of the rifle, and air rifle pellets are too light to give long range accuracy. Even if this problem could be overcome, we should never lose sight of the fact that the most powerful air rifle is still a puny weapon, incapable of killing purely through massive energy and still requiring precision shooting to kill cleanly.

Non-lethal though the air rifle is, it is still capable of causing serious wounds, and gun safety is as important for the air rifle as it is for any other type of weapon. Safety is but a frame of mind - no more, no less - in which one is constantly aware of where the barrel of the weapon is pointing. Beyond this state of mind a few simple rules should always be followed. Never load a gun until you are ready to shoot. Always treat a gun - loaded or not - as though it were loaded. Never put a loaded gun down.

The air rifle is still happily relatively free of restrictive legislation (at the time of writing - you can check on future legislation at the Police Station). Basically unless you have been detained under a custodial sentence recently, then you may have in your possession and shoot an air rifle under certain circumstances. Anyone under the age of fourteen may not handle an air rifle unless supervised by a person over the age of twenty-one. A person aged between fourteen and seventeen may not buy an air rifle or pellets, but may be given them as a gift. Persons over seventeen may buy, own and use air rifles.

It is an offence to discharge an air rifle unless you have the permission of the owner of the land you are on. It is an offence to possess or discharge a loaded air rifle in a public place. An offence is created if a pellet travels past the boundary of the land over which you have permission to shoot. It is an offence to discharge an air rifle within fifty feet of the centre of a carriageway if by doing so you create a nuisance. A serious offence (armed trespass) is created if you take an air rifle, loaded or otherwise, onto private land where you do not have permission to be. Generally, the laws relating to air rifle ownership and usage are straightforward and sensible.

An air rifle may be viewed as something of an investment. The sometimes high inital purchase price is in marked contrast to the low running costs, for at the time of writing 500 pellets cost around £2. A good air rifle will give many years of faithful service and ask nothing in return save the occasional wipe over with an oily rag.

Chapter 2

The Air Rifle

The vermin control air rifle, sight and pellet must conform to the basic need to be able to place the pellet accurately into a very small target area in order to achieve clean kills. It therefore follows that accuracy is of paramount importance in the selection of hardware, and no-where is that requirement stronger than in the choice of a rifle.

Contrary to popular opinion, high power is not a necessity for air rifle vermin control; in fact the accuracy of spring air rifles tends to deteriorate as the power increases - most particularly within any which are altered to achieve higher than designed power levels - it must, for the piston must move more quickly to give higher power, and the action/reaction which turns this extra piston momentum into recoil is a fundamental and inescapable law of nature. The legal power limit for air rifles within the UK is twelve foot pounds, at which a 14 grain .22" pellet will be travelling at 621 feet per second and an 8 grain .177" pellet will be travelling at 821 feet per second. The modern air rifle is usually designed to give velocities of just under these levels and any attempt to increase on them will not only generally cause greater recoil but will also subject the rifle to higher stresses leading to greater wear. Add to this the fact that the airgun pellet is designed to give accuracy within certain velocity limits which don't include very high velocities, and you'll appreciate why power is always secondary to accuracy in importance. If there's still some doubt in your mind, still a nagging feeling that a couple of extra foot pounds would account for extra vermin, then consider the following. A lightweight pellet travelling at 1,100 feet per second gives something like twelve to twenty yards of over the limit performance, after which it will have slowed to the point where it possesses twelve foot pounds energy and so it behaves like a pellet fired from an ordinary air rifle. All the extra recoil and stress on the gun is for so small a gain - and even at point-blank range the high power air rifle (or, for that matter, even a 100+ foot pound rimfire rifle) simply won't kill with a poorly placed shot.

Air rifles break down into clearly definable groups according to their operating method. The main two groups are identified by the way in which the air to power the pellet is compressed. The most common type is the spring air rifle which compresses the air *after* the trigger has been pulled by driving a piston up a cylinder by means of a powerful spring or pre-compressed gas, whilst the other already has the air compressed in a small reservoir before the shot is taken and merely releases this air as a part of the firing cycle. The former group are called Spring air rifles, and the latter Pneumatic air rifles.

Pneumatic air rifles fall into three groups, the oldest of which (in fact, the oldest of all air rifles) is the reservoir type, which has a quite large reservoir into which enough air to power many shots is pumped. These are the air rifles with the highest power potential of all, and some very high power examples were used in warfare and for deer poaching in times gone by! The reservoir rifle tends to be much more expensive than other types, and is often rather heavy as the air cylinder must be strongly made it if is not to explode under the high air pressure. In favour of the reservoir rifle is the fact that it is to all intents and purposes recoilless, and, fitted with a high quality barrel, capable of great accuracy. The second type of pneumatic has a much smaller reservoir which contains just enough air for one shot and which is re-filled with a pump which is built into the rifle. This is the Pump-up rifle, and, like the reservoir rifle, it is virtually recoilless. The drawback with pump-ups is that they are very inefficient at converting the user's energy into pellet energy, and after pumping up most, you'll be too tired to shoot accurately. The opportunity for variable power offered by the pump-up (often quoted as an advantage) is actually a drawback, as you will only achieve accuracy if the velocity and hence the trajectory of the pellet is constant, shot to shot. The last type of pneumatic is the air cartridge, which contains air within a separate cartridge which also holds the pellet. A striker hits the rear of the cartridge to release the air. This rifle is also high-on recoilless, though against it must weigh the facts that the user is limited as to the absolute number of charged cartridges he can carry in the field and that pumping up the air cartridges can be very time consuming.

Spring air rifles are classified according to the method used to compress the mainspring. Some rifles have a lever for the purpose, situated either under the barrel or action (underlever rifles) or to one side of the action (sidelever rifles). The separate cocking lever gives extra weight over the last, and most common type of air rifle, the break barrel, which uses the barrel itself as the lever to compress the mainspring.

Both underlever and sidelever air rifles are again gaining popularity as people realise the benefits of having a fixed barrel in terms of both accuracy and longevity. The sidelever is most useful for prone (lying) shooting, which is an oft-used position for rabbit shooting, as the rifle is easily reloaded in that position. Many people prefer the underlever as the central cocking lever of this rifle gives perfect balance whereas that of the sidelever places a very slight bias to one side.

Break barrel air rifles tend to vary immensely in quality, as the design is used for many cheap and junior rifles with low or immensely variable power and indifferent accuracy. A good example from a well-known manufacturer, however, will possess sufficient accuracy and power for vermin control and will also give many years of service. The best examples feature a manual latch under the barrel which assures perfect barrel re-alignment. The break barrel air rifle is usually (though by no means always) lighter in weight than lever rifles and so is popular with slightly-built users.

All spring air rifles offer certain advantages over the various alternatives. They are very efficient at converting the energy expended by the user in cocking the rifle into pellet power - many times more so than pump-up air rifles. Mass production gives a price advantage over pneumatic rifles which are produced in far lower quantities. The spring air rifle can usually be maintained and even repaired by any competent handyman at home, whereas the complicated valves of pneumatics are not suited to 'home gunsmithing'. The spring air rifle is quick and easy to re-load, and, above all, it is very reliable.

When selecting an air rifle, there is one very important point to bear in mind. It is best to buy a well-known make which offers a ready spares supply, because there is little more frustrating than having to retire an otherwise perfect air rifle for want of some small component. After-sales service varies immensely, from brilliant customer care to 'couldn't-care-less' attitudes.

Obviously, the quality of manufacture of the rifle is a good guide to its likely performance, and this quality will be reflected in and may be judged by the finish of the rifle. Unless you have indoor bench-testing equipment then there is no real way to adjudge the potential of the barrel for accuracy. You can, however, ask to try a few shots with the rifle and sample the trigger action, and as a good trigger is of equal importance to a good barrel for accuracy the trigger will be a good guide to the general likely performance of the rifle. There are just two types of trigger: good triggers and bad triggers. A good trigger is one which allows you to accurately judge the exact point of travel at which the rifle will discharge - a bad trigger won't.

Double-pull triggers are found on many quality air rifles and have two stages of travel. The first stage offers low pressure and terminates in a 'stop' which may be felt through the blade. Any further rearwards travel after this point will cause the rifle to discharge. A single pull trigger does not have the first stage of travel. In either type, look for any 'creep' (trigger blade movement before discharge) and reject rifles with too much, as there is no way of knowing exactly at which point in the travel of the blade such a rifle will discharge. This is important because no-one can hold a rifle perfectly still and it is necessary to know exactly when it will discharge.

Always ask to try a few shots with an air rifle before you buy it and when you shoot try to concentrate on and judge the trigger action. Consider whether the weight and cocking effort of the rifle are within your own capabilities. Also listen for strange noises - any metallic gratings, knockings or other untoward sounds mean that the rifle is not a good example. To be honest, it takes far more than a few shots to know whether you will be able to shoot well with any gun, for you could take a hundred poor shots and suddenly 'click'!

Despite everything already considered with regard to accuracy, the single most important aid to accuracy is personal confidence in the rifle being used. If you feel confident with a rifle then the chances are that you will shoot well with it - the opposite is also true and many perfectly good air rifles are placed to one side simply because the owners have lost confidence in them. Your best route to initial confidence is to buy a well-known make of rifle.

Although very much a secondary consideration, the power of the rifle should be questioned. Don't expect any air rifle to guarantee 11.999 foot pounds, for every manufacturer builds in a safety margin to ensure that the rifle does not stray over the legal limit. A .22" vermin rifle should give a velocity of at least 550 feet per second at the muzzle with 14 grain pellets, and a .177" a velocity of at least 750 feet per second with an 8 grain pellet. These velocities sound quite respectable - the energy levels might not, for the .22 quoted is giving 9.4 foot pounds and the .177" 10 foot pounds. The importance of using velocities rather than energies in assessing an air rifle for vermin control will become apparent later in the book. Remember that these are recommended minimum muzzle velocities and there is a safety margin built-in so that a shot at slightly lower velocity will not carry an unacceptably high risk of wounding.

If you find that you cannot afford to buy a suitable air rifle for vermin control then you have two alternatives. Firstly you could save until you have sufficient funds, and this is recommended. Secondly, you might consider buying second-hand.

One consideration against buying second-hand is the fact that major advances have been made in air rifle technology in recent years, so that today's air rifle will not only out-perform designs considered competent a few years ago, but will out-perform them by a considerable margin. Triggers have improved beyond all recognition. The use of modern materials, such as the PTFE piston seal, gives far more reliable power than was available previously. The modern air rifle has been developed to the same extent as the modern motor car - and like the modern car, it is more efficient and powerful than predecessors.

There are many bargains to be found in used air rifles. Ensuring that you get a bargain and don't buy a 'pup' is no easy matter, for the internals of an apparently sound air rifle could be in a terrible state! The only way to evaluate a used air rifle is to physically test it for power, consistency and accuracy. Even then you have to know not only how to test a rifle but also how to judge the discharge noise which can serve as a guide to internal problems which will later manifest themselves.

Examination of the screw heads on a rifle can tell you a lot about its worth - any sign of burring is a sure sign that the rifle has been stripped by someone who was not equipped to do the job - if someone cannot even undo a screw properly then they will probably have made a terrible mess of the more difficult jobs, such as removing the mainspring! Despite warnings of invalidated guarantees too many air rifle owners insist on stripping their own guns using totally unsuitable tools such as electricians' screwdrivers and the like. Generally such people are trying to gain extra power - fitting massive springs which stress the trigger action and age the gun considerably. Don't buy secondhand if the rifle has been treated thus.

Having looked at the lower end of the market, what of the very top end - the converted and customised rifle? A few years ago there was a very strong case for improving the production air rifle as the performance of some models could vary immensely from example to example. Nowadays things have changed, for air rifles from major manufacturers usually perform to a very acceptable standard. Nevertheless, some people demand absolute perfection in their rifle, and for them, the conversions house still has something to offer. Those who do demand the last word in performance, it must be said, are usually field target shooters - people for whom a miss is a tragedy. The person who needs an air rifle for vermin control need not be nearly so demanding, for there is one major difference between his shooting and that of the field target shot. The field target shooter has for a target a 2" circle at ranges between five and fifty-five or so yards. He has no choice in the matter - he must shoot every target no matter what the range. The person engaged in vermin control can alter his ranges by simply getting his quarry to come close to him or by moving closer to his quarry. He can decide against the long-range shot which the field target shooter must take. The vermin controller's target may be smaller, but he can elect to take only those shots he knows to be 100% within his and his rifle's capabilities.

The Mastersport from Airmasters

However, the fact remains that a top-class converted rifle from a reputable source (such as the Mastersport rifles illustrated) will be capable of greater accuracy than the production rifle; furthermore they are a delight to use.

So whilst there is nothing wrong with having a conversion on the air rifle (apart from invalidating the manufacturer's guarantee - make sure the work and the rifle will carry a guarantee before commissioning any work), it should never be considered a necessity.

(Airguns facing left)
From top: Feinwerkbau Sport fitted tyrolean stock, Feinwerkbau Sport fitted sporting stock, Weihrauch 77 underlever with tyrolean stock, Vulcan custom. All Airmasters.

A Knok-Down field target
The 'kill area' may be reduced with special kits from the manufacturers to make more realistic vermin practice targets.

Chapter 3

Sights

The majority of air rifles come complete with a perfectly good and time-tested sighting system called open sights - which many people nowadays rip off in favour of expensive modern telescopic sights. Open sights comprise the foresight, which is usually a blade or post situated at the muzzle of the rifle, and the rearsight, which is an adjustable cross blade fixed further back along the rifle. The rearsight usually has a 'V' cut into the top edge, and to use the sights the rifle is held in such a position that the foresight tip appears in the middle of the top of the rearsight 'V'. When the rearsight is correctly adjusted for a particular range, the pellet point of impact should be just above the spot indicated by the foresight tip.

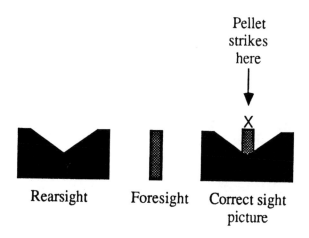

Pellet
strikes
here

Rearsight   Foresight   Correct sight
picture

The rearsight has to be raised in order to raise the pellet point of impact and vice-versa, moved to the right in order to make the pellet fly further to the right and again, vice versa. One particular type of open sight has a rearsight with a tiny hole through which the user looks, and the fact that the hole is so small means that everything from the close foresight to the distant target appears in sharp focus. These sights are known as 'peep' or 'diopter' sights, and are used in the main for target shooting. The severe limitations they place on field of view put their field usefulness in question. Open sights, though, are perfectly adequate for most vermin control applications, and the only time they are really inferior to telescopic sights is when the light falls to a level where the open sights simply cannot be seen.

Many people nowadays seem to regard the telescopic sight as essential for vermin control; they forget that for a century others accounted for vermin using quite basic air rifles and open sights - making up for any deficiency in gun and sight with fieldcraft which enabled them to get within range of their quarry.

In fact, open sights have advantages over the telescopic sight. For a start, they are fitted to most air rifles and so effectively cost nothing! Perhaps less obviously, the sights are far more robust than telescopic sights and far less prone to faults. A telescopic sight fault can only be found by substitution and may cause inaccuracy which is firstly blamed on the rifle or the user - a frustrating waste of time is involved in tracing the fault. Thirdly, the open sights are mounted lower down than a telescopic sight, which gives certain advantages in sight line/trajectory relationship, as will be seen later in the book.

Open sights are perfectly adequate for 95% of vermin shooting, although there are occasions when a telescopic sight is highly desirable.

Telescopic sights have been popular with air rifle hunters for only a few years, but in that short space of time they have made a terrific impact. A telescopic sight is simply a telescope within which is a reticule (cross hairs or post) which appears in the same plane as the target. Using a scope could not be easier - you simply put the cross hairs where you want the pellet to go! Having said which, don't get the idea that you cannot miss with a scope - far from it! Having a degree of magnification makes the target appear much larger but it also magnifies any movement of the rifle, and many newcomers find it initially difficult to get on with using a scope. When the light fades, though, the scope comes into its own, as you can carry on shooting long after the open sight man has given up because he can't see either his sights or the target any longer.

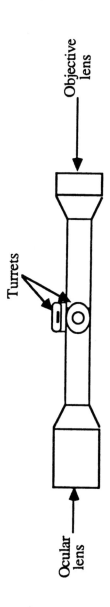

Typical telescopic sight

If you are looking for a scope for vermin shooting, firstly forget any scope which has a body tube of less than 1" diameter - i.e. the cheaper scopes - these are satisfactory for shooting at tin cans in strong daylight but less than perfect for vermin shooting in the evening. 1" tube scopes don't come with mounts as do their cheaper, smaller bretheren, and a reasonable quality scope with mounts will set you back at least £50 - don't settle for less! Scopes are specified according to the diameter of the objective lens in millimetres and the degree of magnification. Hence a 4x40 scope will have four times magnification and an objective lens of 40 millimetres. The 4x40 is the great all-rounder, with the 4x32 coming a close second. Some scopes have variable power, and these are known as 'zoom' scopes. Personally, I prefer a fixed scope as not only is there less to go wrong than with a zoom but you also always see the target at the same degree of magnification and so learn to judge range.

Much psuedo-scientific writing over the years has thoroughly confused the issue of how well individual scopes allow the user to see in low light levels. One oft-quoted measurement is the exit pupil - actually the unobstructed diameter of the objective lens divided by the power of magnification. In a 4x40mm scope this will be 10mm - it actually means that a 10mm diameter circle of light reaches the eye of the viewer - no more, no less. Exit pupil sizes might not always be as you might expect, for the 4x40mm scope used as an example could have internal masking to reduce the exit pupil to far less than 10mm. Bearing in mind that the pupil of your eye, through which the output of the scope must pass, will reach a diameter of 5mm-6mm in low light (after which further growth is meaningless because there simply is not enough light to see well enough to shoot), it will be seen that a large exit pupil is superfluous. Simple mathematics show that in theory a 4x24mm scope will give a 6mm exit pupil and so there is no point to having a larger objective lens to increase the exit pupil. True but for a production detail. The quality of resolution of an image is higher at the centre of a lens than it is at the edges, so that using a 32mm or 40mm four power scope means that you can view the image through the best part of the lens and so see a higher quality image. A really good 4x24mm scope will probably be even better than a 4x40mm in this respect, though the costs of producing near-perfect optics are quite staggering and such a scope would cost many hundreds of pounds.

The ability of a scope to allow you to see through it in low light is determined by optical quality - never on size. One scope used to be advertised with 'luminosity' figures - actually the exit pupil size squared - which looked really impressive. The 4x32 had a luminosity of 64, the 4x40 had 100, and the mighty 45mm zoom a maximum luminosity of 199! In fact, the scopes shared a very similar performance, the only major difference being that the best were the fixed power scopes. Many people have fitted 56mm or even 65mm low quality scopes to their rifles and discovered that not only is the image quality poor for low light use but also that the extra height of the scope above the rifle makes for an alarming sight/trajectory relationship, for the pellet must rise in relation to the sight line at so steep an angle that it flies far above the sight line at all middle ranges - accurate shooting is nigh-on impossible! The large scope has a macho appeal and for those who can resist it finding a good quality scope is still far from an easy matter.

Optics is a vast and complicated subject and rather than join the none-too-exclusive club of airgun writers whose psuedo-scientific jottings have mislead air rifle users I prefer to offer a simple practical comparative test which allows anyone to evaluate scopes for optical quality. In normal daylight it is difficult if not impossible to evaluate scopes beyond identifying the most marked of faults (called 'aberrations'). In low light, however, poor scopes can be identified easily because you won't be able to see clearly through them! Simply try a selection of scopes in low light and it will be immediately obvious which are the best. Try also to view an upright shape, such as a telegraph pole, through the edge of the scope, for a poor scope will 'bend' the image - the same goes for horizontal lines.

Parallax error is another term which seems to cause airgun writers problems. Simply - and in the context of the telescopic sight and not in a photographic sense - parallax error is apparent image movement caused by not looking down the axis of a scope. Simply, if you look to either side of the axis you will still see the target area, but in one case the cross hairs will appear to the right of this and in the other case they will appear to the left. Obviously, there can only be parallax error if the exit pupil is large enough for the pupil of your eye to view from many angles - you won't get parallax error in a scope with a 6mm exit pupil as you must look down the axis of the scope! The effect of parallax error is inaccuracy to the extent to which your eye is away from the axis of the scope. There are several ways to combat parallax error, the first of which would be prevention - making scopes with a 6mm exit pupil. This would be easy with internal masking in prime or fixed power scopes, but a bane in zoom scopes for the following reason.

Imagine a 4-12x40mm zoom scope - its exit pupil varies between 10mm and 3mm! Anyway, rather than internally mask scopes to a 6mm exit pupil, the industry usually fit special objective lens housings which allow the objective lens to be precisely focussed at the desired range. Such scopes are usually called 'parallax adjustable'.

The simplest way to avoid parallax error is to look through the axis of the scope, which entails always mounting the rifle in exactly the same way. This is easiest achieved to an extent by having a stock with a very high comb which positively locates the cheek, although no special stock ensures correct head placement in varied shooting positions - off-hand, sitting, kneeling and prone shooting positions all produce different head/scope relationships! In the final analysis, parallax error is only a real problem if you are an incredibly inconsistent gun-mounter. For all practical purposes, the main advantage of parallax-free shooting is the extra confidence it gives you.

Inside a telescopic sight there will be up to twelve or even more separate lenses which combine to magnify the viewed image. If any one of these lenses moves the slightest out of true then the viewed image also moves. A loose lens which can move as the rifle is discharged will move the viewed image to a new position with every shot and accuracy will be impossible. Thus a telescopic sight must always be treated gently in the field.

Some writers have in the past bestowed their telescopic sights with almost magical properties whereby they allow their lucky owners to see in almost total darkness! The truth of the matter is that although a scope will permit you to continue shooting when light levels are too low for open sight use, there is a point at which no purely optical instrument can enable you to see well enough for vermin shooting. To lay an old myth finally to rest, the telescopic sight will not enable you to see well enough to be able to shoot rabbits over snow even on a fully moon-lit night.

Unlike the air rifle, a scope is a delicate instrument which is quite easily damaged. The outer lenses are the most vulnerable parts. The front lens (the objective lens) can be damaged quite easily in the field and even when the rifle is standing idle in the house that lens tends to gather dust unless the lens caps are fitted. Cleaning off the dust can lead to the lens becoming scratched, and the best equipment to use is a camera cleaning kit, available from most photographic shops.

In addition to open sights and scopes there is another type of sight often feted as useful for the air rifle shooter. This is the 'dot' sight, which I prefer to call the 'binocular' sight as both eyes are used with this type. Basically, the one eye views the sight itself, while the other eye looks directly at the target. The eye viewing the sight does not see the target at all, but only sees a red dot or some similar shape. In use, the brain superimposes the dot viewed by one eye on the target as viewed by the other.

If both eyes are focussed on the target then binocular sights can give accuracy - however, the eyes can move quite independently of each other, so that if the user concentrates for a fraction of a second on the dot then the eye viewing it will move, so that the dot appears to move in relation to the target area. Careful aiming of the type necessary for vermin control is therefore not suited to this type of sighting system and it is not recommended. Do not confuse binocular sights, incidentally, with scopes which have illuminated reticules.

The best advice regarding sighting systems I can give the novice is to start off with open sights and learn to shoot - *then* get a scope.

Chapter 4

Pellets

Like the rifle used in rabbit hunting, a pellet should be chosen above all for accuracy - almost any, if not all, pellets will kill a rabbit outright *if* you can hit it in the right place - there is no magic shape or material which will permit humane killing with poorly-placed shots.

Pellet accuracy is dependent on two basic factors - firstly, consistency of shape and weight - and secondly, it is also dependent on compatibility with the rifle. This may strike some people as strange, but a pellet which gives good accuracy and power in one rifle may give poor accuracy and power in another otherwise powerful and accurate rifle! When selecting a pellet, it's best to start off with a collection of *good* quality makes and then shoot groups off a bench to assess the accuracy of each (see Chapter 5). Then, and then only, get your rifle chronoscope tested with the best two or three types, and go for the best combination of accuracy and power.

The actual type of pellet chosen is not quite so important as the quality of the pellet, though it's worth remembering that pointed types are quite easily damaged, flathead types lose velocity far too quickly, and that a good roundhead pellet offers the best compromise! With quarries where a body shot can reliably kill, then it is sometimes worth harnessing the penetrative qualities of the pointed pellet or the shocking power of the flathead, though the normal target of the brain case requires so little in the way of penetrative powers for a kill that the roundhead gives all that's required in this department.

The question of which pellet to use for vermin control raises the question of which calibre to use. For years, manufacturers have made ninety-five .22's for every five .177's - as everybody 'knew' the .22" was a man's calibre - the only hunting calibre, but nowadays things are not nearly so clear.

The .177" pellet is a real lightweight and travels much faster than a .22", which means that it drops less in flight so that the user has to 'aim off' much less. Remembering that accuracy is the prime consideration, I think that the less you need to 'aim off' the better, and hence the .177" is worthy of consideration.

A .177" rifle giving around 800 feet per second with the selected pellet and scoped will allow the user to aim dead on at all ranges between eight and thirty yards if the rifle is sighted in at twelve yards. This is because the pellet rises to cross the sight line and then carries on flying over the top of the sight line until it comes back down through it at around the twenty-eight yard mark - very useful for all-round vermin work.

In the final analysis, the points to look for in a pellet are quality and consistency - everything else is a secondary consideration. Unusual and novel shapes or materials may not necessarily be an aid to good shooting.

Other equipment

In some vermin control work there is a case for wearing special clothing for concealment purposes when shooting in natural cover. Camouflage, so beloved of many an air rifle hunter, is not strictly necessary - in fact *any* clothing in drab browns and greens will suffice for most woodland or hedgerow camouflage, though if you are, for instance, shooting roosting woodpigeon then don't forget to cover your hands and face as well. This book is about vermin control rather than airgun hunting pure, so that in order to meet the far wider range of shooting areas a range of normal clothes in a variety of hues will suffice. Far better and more versatile than any camouflage pattern is concealment. With camouflage clothing always bear in mind that it only works against a suitable woodland background, and even then only as long as the wearer keeps quite still. Dress from head to foot in camo gear and take a walk through the woods and you'll stand out quite as much as someone wearing denims! Camouflage is a positive liability when seen against a uniform background, such as concrete, corn, grass or even plough. Despite this many people would not consider so much as shooting a tin can in the garden unless they were draped in full camo gear!

A knife is almost a necessity for vermin control, as you are best advised to paunch rabbits as soon as possible after they have cooled a little to avoid the stomach-turning stench which rises from a long-dead rabbit cadaver and the knife can also be used in hide building and for many other purposes. A folding knife will be quite sufficient, and this avoids the 'scare factor' of the fixed blade sheath knife, since it can be carried unobtrusively in your pocket. Always buy a locking knife - otherwise you could find the blade closing on your fingers! The best rule for knives is to buy one with a large handle and a small balde: this gives you real control when you cut, and no-one will view it as an 'offensive weapon' as they will a 'Rambo' knife. Keep the knife very sharp - there's nothing worse than trying to gut a rabbit with a blunt knife - and this will mean regular sharpening sessions, as rabbit bones are quite hard and blunt a knife very quickly. Most people have the manual skill to sharpen a knife with a stone, keeping the necessary 20 degree edge, but there are many small pocketable sharpening devices on the market.

Since any rabbit shot through the brain dances around alarmingly you'll want to manually go through the motions of dispatching it, just in case, and this is a case of do as I say - not as I do, for I am one of the Bruce Lee school of rabbit dispatchers who use the 'rabbit punch' (a blow with the side of the hand against the base of the rabbit's skull where it joins the neck). My advice is to either learn to break the rabbit's neck by pulling it across your knee, or preferably, if you're new to the game and a little squeemish, getting yourself a priest.

A priest is a cosh, weighted with lead, with which you crack the rabbit's head/neck to kill it. This saves bruised hands, and is a lot fairer to the rabbit than a whelter of ineffective blows with the side of the hand.

A bag of some kind can be useful to carry a large number of rabbits, but I prefer to 'string' the rabbits by making a cut behind the tendon of the thigh and threading the other leg through this, enabling me to thread the rabbit onto my belt. The drawback with this is that the blood from the freshly opened gut cavity drips down your leg as you walk!

We are now equipped and ready to go in pursuit of vermin. Before you avidly read on, why not take a little time off and hone your marksmanship? You need to be able to shoot groups of roughly half an inch at normal hunting ranges before you are ready.

Chapter 5

Marksmanship

All of us, whether we are vermin shooters, field target shooters or just plinkers, wish that we could shoot more accurately, yet few people know how to practice in order to improve their standard of marksmanship and most merely push ever more lead in the general direction of a target in the vague hope that their marksmanship will one day, as if by magic, improve as a result.

In the following section, I hope to show how target practice may be structured and the results analysed in order that the individual will be able to recognise exactly where he or she is going wrong - what impediments they have to accurate shooting. Once a fault is recognised it may be corrected, and to keep on shooting in the hope that faults in technique will correct themselves would be as futile as for a Doctor to prescribe the contents of a medicine chest to a patient because he could not diagnose the ailment!

The first consideration when in pursuit of improved marksmanship is whether the air rifle, sight and pellet combined are really capable of top accuracy, as there's little point in practicing with an inaccurate outfit! It's quite possible that your current equipment is letting you down to some extent, for if it is only capable of, say, half inch groups at ten yards then that inaccuracy will be added to your own and leave you with gross inaccuracy! Begin, therefore, by putting your equipment to the test by shooting off a bench rest at a paper target.

To test your equipment, you will need a wind-free range of at least ten yards, and preferably twice that distance. The target should be a simple piece of thick paper with a single small dot marked on it as an aiming point. The bench rest should, ideally, be either of sufficient height for you to shoot comfortably standing, or even better, sitting in a chair, and the whole object of using a bench rest is to take away any inaccuracy on your part so that the equipment itself may be tested.

The gun needs to be rested on a surface which will permit it to recoil in the normal fashion as though it were being shot off-hand, and a cushion or two will suffice for this purpose, although if you can get hold of a proper sandbag then use it, with a little padding on top. Obviously, the cushions or sandbag need to be on a firm surface which does not move at all - flimsy camping or wallpapering tables are useless. The ideal commonly available support is a solid workbench. If you cannot get hold of a solid enough table then build one by driving posts into the ground and nailing a top on them! Remember that the merest movement in the resting surface will ruin the validity of any testing.

Once the range is set up, the rifle should be sighted so that the pellet hits about half an inch under the aiming mark. This is so that the aiming mark is not destroyed as you attempt to shoot your groups - obvious when you think of it! You are now set to test your rifle, sight and pellet combination - and the results could be quite staggeringly good or bad - depending partly on the quality of the equipment and also on the compatibility of the elements of the outfit. You may find, for instance, that your favourite pellet simply does not suit the barrel of your gun whereas another pellet is searingly accurate. Thus you should have a wide selection of pellets available for your testing.

In choosing pellets for testing, firstly consider their intended use - if you want to shoot field target then you need a pellet which loses velocity slowly in flight to lessen the effects of pellet drop at longer ranges, if you wish to hunt then you might require a pellet which transmits the greatest 'wallop' to the quarry and which is not so good at maintaining velocity in the air. Shortlist suitable pellets and buy a selection of these. The more pellets you can try the better, for barrels vary minutely between examples, and even two apparently identical air rifles can favour different pellets.

There are a few points to check before you can really test your outfit. Firstly, make sure that your air rifle is not burning oil when fired by examining the bore for smoke immediately after a shot has been taken. If there is noticable smoke, then 'stringing' of shots in a vertical line on the target could well be caused by this. Check that your scope, if you use one, is above the centreline of the rifle, for if you have the wrong scope mounts fitted so that the scope is to one side of this line, then the rifle will shoot either left or right at all ranges except the range at which it is sighted in. Check that the bore is clean and free from dents, as a small defect in the bore can have a disastrous effect on accuracy. Finally, check the tightness of all screws and bolts on rifle and scope mounts.

You should now be ready to begin accuracy testing your equipment. Pick a day when the winds are slight if you're shooting outdoors, and begin by learning how to shoot from a bench rest. Bench rest shooting gives better accuracy than any style where the user supports the gun himself, yet there is still room for technique improvement. You will notice how easy it is to keep the rifle sighted on target from the bench, so that any improvement in your technique must come in the form of improved trigger pull. Practice pulling the trigger blade straight backwards instead of slightly to one side, as most people do, and concentrate on 'follow-through'.

Follow-through is a term used in target shooting which describes keeping the rifle on target after the shot has been fired, and at least until a second or so after the pellet has struck home. Since most people will be using their scope as a spotting scope merely keep looking through the scope after you've pulled the trigger, and keep the rifle quite still until you've noted the position of the new pellet hole.

When you are happy with your own ability to shoot from a rest, then begin serious pellet compatibility testing, by firing groups with each type of pellet in turn and writing down the results. You can measure the group size easily enough with a pair of dividers and a ruler, and the measurement of interest is the greatest distance between the outer edges of any two pellet holes from the group. In order to find the commonly-quoted 'centre-to-centre' (c-t-c) measurement simply subtract the pellet diameter from this measurement. It's best to include any 'fliers' in the measurements, as a pellet which gives fliers is of little use for either the hunter or the field target shot. Shoot groups with all of your sample pellets, and if they all give you fliers then look elsewhere for the problem - which could be with your rifle, your scope or your technique. Firstly, check whether your scope is at fault by shooting with open sights, and if you can then group you know what action to take! If you still get poor groups with open sights, then check your own marksmanship by asking someone else to shoot the rifle from the bench. If they too get poor groups then your rifle is at fault and in need of attention.

What exactly constitutes 'good' accuracy from the bench? Well, most air rifles using compatible pellets will group better than three-eights of an inch at twenty yards off a bench rest - the best ones will group better than one-fifth of an inch at the same range (both figures c-t-c). Since your groups might, in spite of your practice, be influenced by your own ability, don't scrap your rifle if you cannot do better than, say, half-inch groups at this stage!

The object of the exercise so far has been to check that your equipment is not wildly inaccurate and to find the most compatible pellet. Once this has been done, then you can start to learn to shoot, and you should find that after even one month's practice, your bench rest groups will tighten up quite satisfactorily.

Good rifle shooting is all about one thing - trigger pull. If you can pull a trigger properly then you can shoot, and even if you have other technique faults they pall into insignificance compared with the simple ability to operate a lever!

The trigger is simply a lever which pivots to give a movement in line with the centreline of the rifle - the problems with trigger pull come not because of the trigger itself but because we operate this lever with a finger which pivots in three places, and which is therefore capable of describing three arcs in natural movement, but which is not naturally geared to provide the straight movement necessary to pull a trigger. In order to correctly pull the trigger, the three joints of the finger must work together in a co-ordinated fashion - a movement which must be learnt - and in order to learn this movement, there is a simple short cut which few people seem to take advantage of.

When an air rifle is discharged, it surges backwards and forwards in response to the rapid movements of the piston, and since these movements take place a fraction of a second after the trigger has been pulled, they mask any movement imparted to the rifle by the pulling of the trigger. If you try aiming an unloaded rifle at some mark, and then pull the trigger, then there is no rifle movement to disguise movement generated by your trigger pull, and you can easily see whether you are pulling the rifle up, down, left or right as you pull the trigger! Ten minutes; practice a day with an unloaded rifle is worth more than twenty-four hours' normal target practice, and since you can practice in this manner indoors day or night there is no excuse for not cultivating a decent trigger pull action.

Supplement your dry firing practice with range practice, and you should see a steady improvement in your shooting, though it is quite possible to develop faults on the range which negate the effects of your improved trigger pull. Since live firing practice involved the discharge of the rifle and all the varied movements and noise so generated, a fault might develop which, like a poor trigger pull, is largely masked by the rifle's discharge. The root of most such faults is a phenomenon known as target shyness.

Target shyness is a psychological problem in which it becomes difficult to actually bring the sighting device onto target in a steady manner. Most people have a natural and subconscious reaction to this by jerking the rifle onto target as a part of the firing sequence, i.e.:- as they are beginning to pull the trigger. This in turn leads to a jerking of the trigger, and hey presto - you're back to square one, with a lousy trigger pull action and little hope of accurate shooting. The problem with this jerk is that it is not controlled enough, whereas a steady rifle movement past the aiming mark can be controlled and combined with a good trigger pull.

There are two proven methods of combating target shyness. Archers used to set their sights underneath the aiming mark, so that they did not have to actually bring the sight to the difficult area, although I favour the alternative method of going back to the bench rest, where the rifle is so easy to keep steady that it may be brought to the correct aiming position.

A tendency to jerk the rifle onto target can also come from poor trigger pull, as if you tend to pull the rifle, say, to the left when you pull the trigger then sooner or later you subconsciously begin to jerk the rifle in the opposite direction to compensate. The cure, again, is to go back to the bench or to resume dry firing in order to again cultivate a good trigger pull.

Shooting an air rifle involves learning certain manual skills, and psychologists have studied these learned skills and know the best ways to acquire them. Any manual skill is best acquired through a 'little and often' regime, rather than through protracted sessions. It has been discovered that a few minute's practice, followed by a break during which a process known as 'reminiscence' takes place, followed by more practice, is the very best way to acquire manual skills. This holds true for airgun shooting. Firing twenty shots and then resting is far better practice than firing fifty or a hundred shots, as the latter tires the operative muscles and performance will deteriorate accordingly, and the person concerned will also psychologically deteriorate - so that further shooting might be described as 'banging one's head against a brick wall'!

Before blazing away at a target, then, you should put yourself in a relaxed frame of mind and decide how many shots you are going to take before relaxing again. Twenty is not a bad number, as repeated re-loading of a spring air rifle plus the fatigue caused by having to support the rifle will rapidly take their toll after more than twenty or so shots. Set up a paper target with four small dots as aiming marks, and fire five shots at each, stopping and thinking about each shot after it has been taken. If you can go through a shot in your mind - rather like a television 'action replay' then it gives you a great opportunity to analyse any small faults in the shot. Having analysed any technique fault in the previous shot, don't dwell upon it too much unless it is a frequent fault, but approach the next shot with a fresh mind as though it was the first of the session. Sooner or later, any regular fault will become obvious, and then you can take action to remedy it.

Your paper target is the best guide to shooting technique faults, and with practice, you will be able to read the relevant fault from the position of the pellet hole. Don't be too concerned with individual fliers, but note general trends, such as several shots flying to the left of the aiming mark. Check that the fault is not due to the equipment or the wind, and then try dry firing to see why you are pulling the rifle off target. You will in time learn to identify your own most recurrent faults, and eventually you will immediately recognise them whenever they occur and be in a position to remedy them.

Marksmanship is one of those awful things where you can

try too hard and your performance always deteriorates as a result of doing so. If you can spend half an hour a day of five minute's gentle shooting followed by five minute's rest, with dry firing or bench practice as necessary, then within three or four weeks your shooting should improve drastically.

The time is now ripe to increase our versatility as all-round shooters by learning to shoot well in various positions, and since shots in the field or on the field target course rarely allow the luxury of a favoured position it is necessary to be able to shoot from the most contorted positions on occasions.

The three basic shooting positions are off-hand (standing), kneeling or sitting, and prone (lying down). Within those positions there are many other positions which may be forced upon us in practical shooting. The most difficult position of all to master is off-hand, and so we'll begin by examining this.

Watch a complete novice shooting an airgun and as often as not he will try to adopt a target shooting stance with his leading elbow tucked into his hip and his body leaning backwards to counteract the weight of the rifle. This position is great for a heavy recoilless target rifle indoors, but in the field it is far too restrictive, and the slightest variation in the actual rifle support forced by localised conditions such as having to stand on a slope will send the pellet way off target. No, the sporting stance must be mastered, however uncomfortable it at first appears. In the sporting style, the arm which supports the rifle must be free from the body, and the upper body angled if anything slightly forwards, so that the shooter leans into the shot. This position allows the rifle to recoil naturally irrespective of slight stance differences - essential in both hunting and field target conditions.

Minor points such as where to grasp the fore-end, exactly where in the shoulder to place the butt, and so on, will vary from shooter to shooter according to his own build and the dimensions of the rifle. The same holds true for other shooting positions, as well, and it is best, in fact, for the individual to try slightly varying his position until he finds one that is comfortable. Some target shooting disciplines tend to regiment the three shooting positions, which is fine if you're on a level range but too restrictive in field conditions. Once you have found comfortable positions then you can begin to step up your practice, and to extend the length of each shooting period. Whereas previously you were constantly shooting from one position which placed a great strain on a few muscles, you will now be utilising different sets of muscles according to your shooting position, and by changing position you can relax some muscle groups and so keep on shooting for longer. Added to this, you will now be at the stage of having acquired the basics and being able to stress the operative muscles without necessarily spoiling your performance - so try shooting yourself into exhaustion from time to time.

So far, we have considered shooting at just one fixed range, and the next step is to gain an understanding of pellet trajectory and its relationship to the sight line in order that you will be able to shoot well at varying ranges by aiming off to allow for the position of the pellet relative to the sight line.

The pellet in flight is unsupported and so it will drop due to the force of gravity, and the longer the pellet takes to reach the target (the further away the target is) the more the pellet will drop. The barrel is not, therefore, aimed straight at a target, but above it to some extent to allow for the pellet drop. This simple concept is complicated by the fact that the sight line (the line from your eye to the target) is situated above the bore line, so that in order to hit the point indicated by your sighting device the pellet must rise anyway.

Let's look at a pellet trajectory in relation to the sight line in its simplest sense. Suppose, for sake of argument, that the rifle is sighted-in at fifteen yards. The pellet begins its flight from a position of perhaps one and a half inches under the sight line, and during the first fifteen yards of flight it is climbing to reach the sight line as the barrel has been aimed upwards. By fifteen yards it has reached the sight line, and thereafter it continues to climb above the sight line to some extent, until dropping to again cross the sight line further out. After this, the pellet drops relative to the sight line.

The pellet drop is affected to a large degree by the pellet's velocity, since, as we have already considered, a slower pellet takes longer to reach the target and hence drops more in flight. This, for a given level of power, makes the .177" pellet far flatter-flying than the .22", which explains why many air rifle hunters nowadays use the .177".

Let's examine a typical trajectory. Suppose that a .177" pellet leaves a muzzle at 800 feet per second. It's velocity will decay at around 10% every ten yards, so that by the time it is ten yards out it will be travelling at a speed of 720 feet per second. The average velocity is 760 fps, and it has travelled thirty feet. From this we find that the pellet took .03947 seconds to reach the target. In order to work out how much the pellet will have dropped due to gravity, we square the flight time and multiply it by 192, which gives drop in inches. So our pellet has dropped by .299" (approximately) at ten yards. You can work out the drop at any range in this manner. At forty yards the velocity will be .9x.9x9x.9x800 fps, or 524 fps. Add this to the 800fps and divide by two and you've got the average velocity of 662fps. The pellet has travelled 120 feet and so it will have been in flight for .18 seconds, and the drop will be 6.22". You can prepare a trajectory graph for yourself in this manner, given the initial velocity of the pellet and its decay rate, and by then drawing in the sight line you can easily see where the pellet will be relative to the sight line at various distances. The computer lends itself to this kind of work, as it is able to perform all the necessary calculations in a fraction of a second and give a complete print-out. By altering the sight line to simulate sighting in the rifle at different distances, you can arrive at the 'perfect' sighting-in range - the one which has the pellet's flight path very close to the sight line at normal shooting distances. Using a .177" rifle producing 850 fps, for instance, you can sight in at around twelve yards and the pellet will be within $3/8$ths of an inch either above or below the sight line at all ranges between seven yards and thirty-three yards, so that you can aim dead-on at any target and be within $3/8$th of an inch!

The method described above for calculating pellet drop is not exact, for the velocity decay varies according to the velocity anyway, but it will be within half an inch at forty yards, and so serves as a very good guide.

You can dispense with theory and find your pellet/sight line relationship in a more practical manner by sighting-in at a certain range and then seeing how high or low the pellet strikes at different ranges. You will need paper targets and a bench rest for this, and if you carry out both theoretical and practical tests, you will be amazed to find that the results agree!

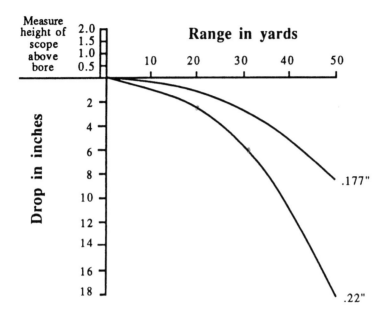

Measure height of scope above bore

Range in yards

Drop in inches

.177"

.22"

If the foregoing sounds like unnecessary hard work, then I suggest that you sight in your .177" at twelve yards, or your .22" at eight yards. Find the second crossing point which this gives (roughly thirty yards for the.177" and twenty-five yards for the .22") and then find out how high the pellet flies at mid-range (around $^3/_8$ths of an inch). Remember that targets closer than the first pellet crossing point will find the pellet still on the way up, and aim over the intended point of impact, that targets between the two crossing points will find the pellet over the indicated point of impact, and aim slightly low, and that target beyond the second crossing point will find the pellet steadily dropping under the sight line, so aim high again.

Once you have grasped the principles of trajectory and its relationship to the sight line at distance, and you have an understanding of where your outfit is shooting at varying ranges, then begin shooting at all sorts of ranges to gain some experience of range judgement. You can learn to judge range best by thinking in terms of paces, rather than yards or metres, as if you do this then you have a built-in measuring system! As you go about your day to day activities try judging the ranges to any objects you will have to walk past, and see how closely you can judge those ranges. Soon you will be able to quite accurately judge range and combining this ability with a knowledge of trajectory will make you an all-round shot.

Chapter 6

Rabbit Control Methods and Tactics

The rabbit is not a native of the U.K., though he has been with us for so long that he is an integral part of the British landscape. Whilst few would wish to see this rodent eradicated, localised populations in favourable conditions can breed quickly enough to present a real problem in damage to crops and, during the winter, to trees. A healthy doe can produce 20 or so off-spring a year, each of which will be capable of breeding at the age of four months. An infestation can therefore arise within less than a year from a single buck serving his many does.

Rabbit numbers are contained quite naturally by cold weather and starvation, by dogs and cats, buzzards and foxes. The rabbit is one of those creatures which counters a high mortality rate with a super-high breeding rate. The garden and modern agriculture create immensely favourable living conditions for the rabbit which reduce natural mortality and so natural population control has to be supplemented with our own.

Rabbits are culled by trapping, gassing, snaring, long-netting, ferreting and shooting. In many circumstances shooting with an air rifle can be the most effective and most, if not only, appropriate method.

It is impossible to think of the rabbit without considering myxomatosis. This disease originally affected the South American rabbit - a close cousin of the European rabbit - and was identified and firstly isolated by a French doctor who infected the rabbits in his garden with it. The effect was electric, for myxomatosis killed 199 out of every 200 rabbits. The disease soon found its way across the channel in 1953/4, and was willingly spread by people who had rabbit problems. Over the years the rabbit has acquired something of an immunity to myxy and whilst the virus mutates and from time to time manages to kill large numbers of rabbits in a new form, the rabbit is slowly acquiring an immunity. Eventually, the disease will be as deadly to the rabbit as the common cold is to humans.

Myxomatosised rabbits are easily identified. The eyes, ears, nose and anus swell and emit a horrible pus - in the later stages the animal is blinded because the eyes swell to the point where they close up altogether. Eating a myxamotised rabbit cannot harm you, though I have yet to meet the person who can face the prospect.

If you have a few rabbits which cause damage in a garden then air rifle control need consist of nothing more difficult or arduous than leaning from a window and shooting them! The majority of rabbit shooting, however, is a far more complicated business demanding the ultimate in hunting and shooting skills.

The rabbit is at the upper size limit of creatures which may be taken humanely with the air rifle. It is a powerful animal, capable of moving at speed even when quite badly wounded, and if it is to be culled with the air rifle then the pellet must be placed quite precisely in the brain of the animal. Don't be tempted to try for the lucky body shot, as even a body shot with a powerful rimfire rifle can wound rather than kill.

Because of the need for very high accuracy in rabbit shooting, the ranges at which shots may be taken are severely restricted. Most people should keep ranges down around the twenty to twenty-five yards mark, though better shots might stretch that maximum range out to forty yards - after which, the limitations of the air rifle pellet at long range make themselves felt. The great problem is that the newcomer, whose marksmanship is likely to be suspect, has the greatest need to get really close to his rabbit, and it is getting close to the rabbit which is the most difficult skill of all to master! Catch 22!

Getting within air rifle range of a rabbit is a task so difficult at times that even the most experienced hunters end up totally frustrated, and in order that the hunter should be in close proximity to the rabbit then it naturally follows that a) the hunter must go to the rabbit or b) the rabbit must go to the hunter. The former is called 'stalking' the rabbit, and this is the most popular hunting method. Not so glamorous, but far more effective under many circumstances is letting the rabbits come to you, an activity known as 'ambushing'. We'll study these two basic methods in turn, starting, as should the novice, with ambushing.

Ambushing

Ambushing has a number of advantages over stalking. Firstly, it is easier, and secondly, because you know beforehand the likely ranges you can zero your rifle precisely according to the particular circumstances (uphill, downhill, crossing wind etc.).

Now, it's no use whatever just sitting down in the countryside and waiting for a rabbit to come innocently tripping along to fall to your trusty rifle. You have to firstly do some detective work and decide exactly, and I do mean exactly, where the ambush is to take place. Let's begin at the beginning. Rabbits live (generally, but by no means always) underground, in burrows - a collection of which is referred to as a *warren*. The animal being basically nocturnal (active at night), its daytime activities tend to be restricted to the burrow or its immediate vicinity. The first job, therefore, is to locate a likely looking warren. Whether it is well inhabited will be discernable from the quantity of droppings near to it. Like many male animals, the buck rabbit marks out his territory with his droppings and urine to keep other bucks away, and I suppose that if we did the same, no-one would ever visit us, either! Fresh droppings, or pellets, are always quite moist - shrivelled-up things resembling currants are far from fresh. Don't incidentally, make the same mistake as a few novice hunters of my acquaintance did by counting larger sheep droppings when deciding on rabbit population levels!

If you've found a warren which looks promising, then your next step is to visit it just before dawn or a couple of hours before sunset. These times represent the starts of peak daytime rabbit activity, and this next piece of advice is going to sound a little strange (I expect many people to ignore it) - sit yourself down downwind of the warren and wait and watch. "What!" you exclaim "*watch* the rabbits!". Yes watch and learn. One hour spent watching your quarry is worth several week's worth of hunting in terms of how much you will learn. While you're still reeling from the suggestion that you merely view the conies rather than commit bunnicide there and then stop to consider that if you take a shot at the first rabbit you see, what have you learnt about the rabbit? Nothing.

Watching the warren brings a great mental calm, unlike waiting for a shot, which really get you het up. Firstly the young of the warren, the kittens, will probably emerge, the first one visible initially only by its twitching nose. Slowly the kitten will raise its nose to catch a little more of the scent wafting past its burrow. Suddenly, so suddenly you may not even see the movement, the kitten will be up and out of the burrow - still very close and ready to dive back down at the slightest provocation. The kitten will sniff the air carefully, sifting through the various scents for any sign of danger, for even at its tender age it has all the nervous energy so apparent in all rabbits. When it is satisfied that all is safe, then it may stray a few feet from the burrow, and soon other kittens may join in. If this is all happening at dusk then it will not be long before a doe is out - the same cautious entry as the kittens, though after a short while she may well suddenly make off strongly away from the burrow. *Lesson One*. Rabbits emerge very carefully, but once on the move, they travel quickly. Watch the doe until she reaches a hedge and again, she will slow right down, and go through the gap very carefully with nose twitching for signs of danger.

Eventually, the buck might show, and this is a good time to learn how to recognise the buck. He will skit around, watering the ground and leaving pellets while he marks out his territory. He looks a little different from the does, as well, though I find it difficult to differentiate in normal circumstances by looks alone - behaviour is a far more certain give-away. For the record, the doe is smaller than the buck and has a narrower head.

Let's examine the evening's rabbit watching in retrospect. Firstly we have leant that it's far easier to get a shot at a kitten - a fairly useless exercise unless the rabbits *really* are eating you out of house and home - and still quite easy to get a shot at the often still milky (feeding young) doe. The kittens appear outside the warren firstly when they are three weeks old and they should be weaned a week afterwards. Getting a shot at the buck is a completely different proposition. There's no glory in shooting kittens or pregnant or milky does - do it only in the severest cases of rabbit damage. In late summer, a wait outside a good warren will give you plenty of shots at three-quarters grown rabbits which do have value as food and which will be capable of considerable damage to crops - particularly during the following winter, as the snows will force them to bark trees and kill them.

You sat downwind of the warren as the rabbits have an acute sense of smell, and if the wind blows from you to them then you'll not see a rabbit. As well as a good sense of smell, the rabbit has very good hearing, as you will find out when you first try to stalk one, and its eyesight, thought by many to be poor, is quite good enough for the rabbit to be able to see a moving human figure at over 100 yards! In addition to this, one of the methods by which rabbits signal to each other that danger is present is to stamp on the ground with their powerful hind legs, and the vibrations in the ground from a human frame is ample warning to all the rabbits in the area. This is why ambushing is so much more productive than stalking, as the wind can take care of your scent, and since you are not mobile the rabbit cannot hear you and is quite unlikely to see you.

When you come to actually shoot at ambushed rabbits, you'll find that each shot empties the area of rabbits. Stay put - after twenty minutes or so you should be rewarded with another shot.

Stalking

Every macho who ever held an airgun must have dreamt about stalking rabbits with it, for stalking is the glamorous aspect of our sport - hunting in its purist form. Stalking is also an art which is difficult to master, requiring a lot of patience, and so it does have the plus point of separating the mature from the immature, so ensuring that the more 'cowboy' element don't stay the course.

We've already established that the rabbit possesses several highly-acute senses which can warn it of the approach or even the presence of a predator, and perhaps it will help you to appreciate how these senses operate by putting the effective range of each into perspective. Imagine that the rabbit has a number of concentric circles around it, rather like the radar defences of a warship. The outer circle, which is over 100 yards radius, is the vision of the rabbit. Many people dismiss the rabbit as having poor eyesight, but to do so is a mistake, for a rabbit can see a moving human being at considerable range. The second circle is the rabbit's hearing, which is much more highly developed than our own. Whispered instructions to a hunting companion in the vicinity of rabbits will ensure that you don't have much success. The smallest circle surrounding the rabbit is its ability to sense vibrations carried through the ground - stomp around and you'll suffer! In addition to these circles, there's the rabbit's sense of smell, which naturally picks up only those scents which the wind carries to it.

So, in approaching the rabbit we have to remember the rabbit's early warning systems and ensure that they don't register us. First and foremost, that means staying out of sight, and although this may seem to pose an insurmountable problem we have the advantage of stalking a creature whose eye is within two to four inches of the ground - imagine trying to stalk a giraffe for a moment and you'll appreciate how much the rabbit's low aspect means to us. Over even the least undulating of ground, by keeping near to the ground by crawling, we can hide behind the smallest bumps, including molehills! Crawling around in pasture frequented by cattle presents its own problems, but there's no need to enlarge on that, for you can walk slowly *through* the herd without the rabbits identifying you. Any other cover, such as bushes, trees or even parked farm vehicles can be used to prevent the rabbit spotting you.

The rabbit's use of its incredible hearing can be witnessed. Watch a feeding rabbit and you'll see its ears swivelling around from time to time like radar scanners. This complicates an approach near hedges or trees where the ground can be scattered with brittle twigs which seem to attract your feet to them. Keeping bunches Of keys in the wrong pockets will also make quite enough noise to scare rabbits. Before setting off on a hunt, take a few steps around in your complete outfit and listen very carefully for any give-away noises. It's surprising just how noisy a wellington slapping against your calf can be, and some clothing rustles like hell. Find out before you get within the rabbit's hearing.

Because of the rabbit's ability to sense vibrations through the ground, pumps are always preferable to wellingtons, even though you will often end up walking in wet and soggy pumps from the morning dew. There's also a technique to putting each foot down to be learnt. Tread down gently onto the balls of your feet - not the heels.

The sense of smell of the rabbit presents more problems than all its other senses put together, because the scent is carried by the wind, which can eddy around in directions you cannot begin to predict. There's a wood near to my house which is on a slope, and when a south-westerly wind is blowing then it does not matter where I enter the wood because a current just takes my scent right around the wood! Over more open and level ground the effects of the wind become more predictable, though the wind is a tricky customer which catches you out when you least suspect it!

Stalking requires skills which cannot be taught through the pages of a book, but which may only be learnt through field experience. While you're out there learning bear these simple rules in mind and you'll learn much more quickly.

1. Keep out of sight of the rabbits until you're ready to shoot.
2. Be quiet during your approach.
3. Approach from down-wind.
4. Learn from your mistakes.

Lamping

The rabbit is generally a nocturnal feeder, and although during the daytime rabbits can be found outside the warrens - especially at dawn and dusk - the real feeding activity takes place at night, when the rabbits will travel far from their burrows in search of food. They follow established routes by scent on the darkest of nights, and these are the very best lamping nights. Let's just pause a minute to shatter an old myth regarding hunting rabbits at night before looking in detail at lamping.

Many people have written about thrilling moonlit rabbit shooting escapades, basing their story - for fiction it is - on a knowledge of such matters gleaned solely from 'The Norfolk Poacher' - an inaccurate though entertaining song about a poacher who ventured out on the lightest of nights. Such stories are best disproved with a practical experiment. When the next full moon occurs, go out in pursuit of rabbits, and I don't care how bright it is or how big your telescopic sight is the fact will emerge that whilst the rabbits can see you and scatter you won't be able to see them! You'll hear them running almost over your feet in an escape attempt, but you won't see them.

People who actually hunt rabbits at night, as opposed to those who dream of doing so from the warmth of their beds, use a lamp and provide their own light. The effect of shining a light on a rabbit is two-fold. Firstly, it illuminates the rabbit so that you can see it. Secondly, if often confuses the rabbit - particularly if you've the foresight to approach it from its escape route side, and the result is often a rabbit squatting in the beam in the hope of going un-noticed. You can walk so close to the rabbit that the shot becomes no more than an easy formality. Perhaps I'm making lamping seem too easy, so it's worth pointing out that there are many problems associated with the sport's successful execution. Firstly, although the rabbit has lost its sight, so to speak, it still has its senses of smell, hearing and touch, and if you blunder about upwind of the rabbits then you won't get to see any! In spite of the difficulties, a practiced lamper is a very effective rabbit controller.

Many people prefer to lamp in company with a friend, with one holding the lamp and carrying the rabbits whilst the other shoots. I prefer to lamp solo as one person makes a lot less noise than two, though the solo lamper has the problem of holding a lamp on target while he takes aim. You can mount a lamp on the top of your scope using a spare mount, or even onto the barrel of your rifle. I normally rest my rifle on the lamp, which is carried in my left hand.

Unlike the lurcherman, who might have to illuminate a course after rabbit lasting a couple of hundred yards, you will need to illuminate only forty or so yards, and if you use a good quality scope the average hand torch will suffice for this. My torch is wired to a rechargeable battery which I carry in my pocket. I prefer the gel types over the old lead-acid ones which can burn large holes in your clothing when the inevitable spillage occurs. Motorcycle batteries cost £10 to

£20, and will last a long time if looked after.

Lastly, don't forget to take a large bag with you, just in case you have a really good night!

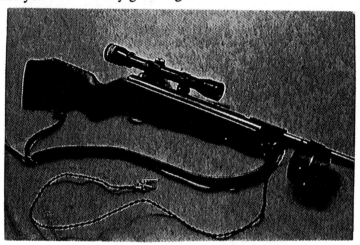

2 photographs of gun
The author's 'home-brewed' lamping outfit.

Chapter 7

Avian Species - Introduction

Modern intensive farming techniques and even human habitation developments can combine to create hugely favourable conditions for certain species of bird. The massive fields of corn, rape, greens and other crops are like larders for woodpigeon and some other birds, offering an ensured food supply capable of sustaining very high population levels at certain times of year. In the winter, bird overpopulations tend to sort themselves out through migration or death through starvation, though during the rest of the year, some form of control is often essential. The crow family can pose other problems. The magpie has in recent years moved into urban environments in huge numbers and like most other members of this family it is a nest-robber, killing chicks and stealing the eggs of small birds. On the game shoot the crow family nearly all constitute a threat to the game birds through their nest-robbing activities.

Whereas in rabbit shooting the target is the rabbit's cranium, or brain case, in avian shooting the target can be either the head or the heart/lung region - the latter not only presenting a much larger target but one which has different requirements of the pellet once it has penetrated.

If a pellet has sufficient energy to enter the rabbit's cranium then it will have sufficient energy to disperse to ensure a clean kill, but when the target is the heart/lung area of a bird the pellet has to be able to disperse much energy in order to kill cleanly. The heavier .22" pellet with its larger frontal area, provided it can penetrate that vital area, will be far more efficient at transferring its energy than the lightweight .177" - and is thus to be considered preferable from this viewpoint.

The major advantage of the .177" for rabbiting is its flatter trajectory when compared to that of the slower .22". In avian shooting, however, most shots will be to some extent uphill into trees - and trees don't grow that tall! The author has been told tales of many successful sixty yard shots at woodpigeon in trees by a hunter sitting a short distance away from base, and a visit to the tree concerned with the 'long-range specialist' has always resulted in the truth concerning the actual range of the shot coming out - and it is normally between twenty and twenty-five yards!

All those characteristics which make the .177" ideal for rabbiting can thus be considered of questionable desirability for avian sport. The main requirement of the rifle is to power a heavy pellet with sufficient force for it to enter the chest of the quarry and do considerable tissue damage once there. One further characteristic of the .22" in the author's opinion tips the balance firmly in the favour of the larger calibre: it is to do with the type of force which the larger pellet possesses.

A .177" pellet penetrates easily, due in part to its small surface area (which, incidentally, also often allows the pellet to emerge from the far side of the quarries' body so wasting valuable force). The .22", on the other hand, tends to penetrate less - though sufficiently to penetrate the vital area - and so its energy is effectively deployed against the vulnerable target area.

When seeking a pigeon and crow air rifle it is important to find one which suits you rather than buying the most fashionable piece of ironmongery of the day. The rifle should be lightweight enough for you to be able to carry it for extended periods, as you might on occasions have to hold the rifle mounted for some time when shooting woodpigeon where there is little cover and you remain undetected by keeping as still as possible.

The open sights which are a standard fitment on the vast majority of air rifles are quite good enough for avian hunting in most situations, provided that they allow the user to see the quarry fairly clearly and don't mask off a large part of it.

There are many occasions, however, when a telescopic sight is a very useful piece of equipment indeed. When the light begins to fail at dusk (just as the birds are coming in to roost) then the scope allows the user to carry on shooting long after the person with open sights has been forced to call it a day because he can no longer see his sights. If the user has any defects in his eyesight then a scope is a great boon at all times of the day because it places the reticule in the same visual plane as the target, so that the user has to focus at just the one distance, and with an enhanced image through magnification as well.

Let's consider which pellet would be most useful for the .177" air rifle owner who wishes to pursue avian quarry with perhaps his number one or only rifle. The .177" has notable penetrative qualities due to its small cross sectional area, and in the case of the chest shot on avian quarry we seek to limit pellet penetration as far as possible. Pointed pellets and other more fancy types of ammunition for which high penetration is claimed are to be avoided, as unless they happen to strike the spine or wing bone of the bird they will merely make a neat hole through its body and not kill it - the round or flat headed pellet is far better. There are two roundhead pellets on the market which offer a quite high weight for a .177" -8 to 9.5 grains - and these will kill birds quite well. The alternative is a heavyish flathead pellet, although it must be said that some of these will not penetrate sufficiently to reach the vital area. The best recommendation the author can thus make is for the long and heavy roundhead pellet.

Even the heavy but high-penetration .22" can pass through the body of a bird, and although the pellet might pass through the bird's heart, it can still leave a sturdy woodpigeon in a fit condition to fly some considerable distance before expiring. Thus again the roundhead pellet remains the theoretical best and, in the author's experience, the best in practice. Heavier flathead pellets will also suffice, although they are not quite so reliable as the roundhead. Some will disagree with this - it's based only on experience.

Chapter 8

Pigeons and the Crow Family

Feral Pigeon

What is a feral pigeon? The term 'feral' describes any domesticated creature which return to live in the wild, and the off-spring of such escapees or releasees. It is unfortunate that, unlike the woodpigeon, the feral pigeon comes in all shades and colours, and so it can be difficult for the untrained eye to differentiate between a feral and a highly-prized racing pigeon. The racing pigeon is in law regarded as someone's personal pet, just as their cat or dog is, and to shoot someone else's pet is illegal. If, however, racing pigeon are damaging crops then they may be shot, although in practice the racing pigeon is so sleek a bird that scrutiny should identify it, and it should not be flocking with ferals. Should you accidentally shoot a crop-raiding bird which turns out to have a ring on its leg, then you should send the ring to your nearest pigeon racing club.

The feral pigeon sometimes flocks up with the larger woodpigeon to raid crops in vast numbers, although the main call for the demise of the birds is usually within towns - especially around factories where the birds will roost at night to take advantage of the heat dissipated from the building - their droppings not only being an eyesore but also damaging to many building materials in the long term.

The first thing to remember if you are asked to deal with feral pigeon around a factory premises is that in order to shoot an airgun anywhere you must have permission of either the landowner or the holder of the shooting rights - and since not many factory premises have rights included in the leases it normally means dealing with the boss - the owner of the factory or of the land itself.

If, therefore, a factory cleaner asks you to go and shoot the pigeons around the factory decline unless you can personally see the boss and ensure that you will be acting legally. A further legal consideration is that an offence is created if any pellet passes beyond the boundaries of the land over which you have permission to shoot. It is necessary, therefore, to make doubly sure that every shot has a safe and very effective backstop to avoid problems.

Shooting feral pigeon around a factory is by no means a sport - it is best considered a 'cull' - a premeditated assault on birds which often make no attempt to leave the site but simply shuffle around as their fellows fall around them.

The air rifle is so effective a tool for the control of feral pigeon that most pest control services utilise air rifles as a part of their armoury.

The best *modus operandi* for a first sortie after feral pigeon is to survey the site and its surroundings in the morning or afternoon on a public holiday when the factory is shut down. Look for perches, which will be easily found because the ground underneath them will be white with droppings! Having established where the perches are, then consider whether each will permit a safe shot, and if so, from what angle? Be ultra-careful in your pre-selection of safe shooting angles, as the consequences of a stray or ricochet shot going 'over the wall' so to speak, are great. You could end up with a fine and the loss of your air rifle.

You will build up a mental plan of the premises, the likely safe shots and the safe shooting positions. You can then plan a route which takes you round these places in an orderly fashion, so that when you've shot one mark, you move on to the next and don't return to the first mark until some time later when the birds have returned. Industrial premises tend to house ferals at most times of day, although your best time is dusk when the birds return to roost.

The feral pigeon is far less robust and much smaller that its cousin the woodpigeon and it can be reliably killed with far less energy than is required for the woodpigeon. If the feral pigeon to be pursued in a built-up area it is advisable to use the softest pellet irrespective of calibre so that the risk of ricochets is reduced to almost nil - the pellets will still kill reliably. The author has often culled these birds within a carpet factory using a 10 foot pound air rifle feeding soft flat-head .177" pellets.

Woodpigeon

Let's kick off by clearing up a minor misconception - the air rifle is useless for protecting crops against flocks of woodpigeon. Magazine writers have long advised the novice to visit farmers whose crops are besieged by woodpigeon in order to gain permission to shoot, but what the farmer wants is the birds harried and dispersed in such circumstances, and that means getting a noisy shotgunner or two on his land rather than a quiet airgunner.

Crop protection apart, every farmer in the land loves to see dead woodpigeon at any time of the year, and the airgunner who can gain shooting permission over mixed farmland should be able to warm the cockles of the farmer's heart with a constant stream of dead woodies taken in a variety of ways.

Roost shooting is the easiest way to kill woodpigeon with an air rifle, and consists of visiting the roosting woods at certain times of day when the birds can be reasonably expected to be in residence. Early and late morning, and late afternoon/early evening (depending on the time of year and available light) give the best times. Early in the morning, the birds leave their night roosts, but often stop off in another wood or copice en route to their feeding grounds. At such times, the birds seem particularly dozy, and will happily fly in without their usual cautious approach, allowing the concealed airgunner the opportunity to account for them in good numbers in a short space of time.

Late morning the pigeon are wont to return to roost to digest the food which will by now pack their crops, and again the concealed airgunner can account for birds as long as he can place his pellet between the bulging crop which acts like a bullet-proof vest and the gut - too high or too low a shot will wound at such times.

There are a few simple rules to observe during roost shooting. Firstly, concealment must be effective as these are amongst the most sharp-eyed of birds.

Secondly, if there is any wind at all at tree-top level then the chances are that any visiting woodie will land into the wind. This is because the woodpigeon is a heavy bird and needs to increase lift by effectively increasing its air speed whilst keeping its speed of approach as low as possible. The airgunner is best, therefore, to sit himself up-wind of the chosen tree, so that the wind comes from behind him. The pigeon will land facing the shooter, presenting a frontal shot.

The third simple rule of roost shooting is to raise the loaded rifle as the bird is making its final approach to the tree - as its concentration is focussed on the branch. For a second or so, the bird will be oblivious to your movement, allowing you to shoulder the rifle - raise it too soon or too late, however, and the wary pigeon will spot the movement and fly off.

Let's put some meat on the bare bones of the three rules above. Concealment is a 'must' when the pigeon is the quarry, and my dislike of gimmickry does not extend to full camouflage for roost shooting. A hide is a godsend if you can build one, but otherwise it is wise to invest in a camouflaged jacket and to wear some form of veil or mask, gloves and trousers in subdued greens or browns. The best camouflage in the world is useless if the wearer moves, however, and hence the advice to raise the rifle at the critical pigeon landing time.

When looking for a good roosting tree or trees to sit upwind of, a little research will indicate the best bets - look for droppings on the ground under favoured trees, and even better is to observe the area through binoculars prior to getting there to see which trees the birds use most.

Try and visualise an arc of fire which gives the potential greatest number of shots by including the greatest number of likely trees, remembering that one good sitty tree is worth several dozen useless ones! A well-prepared airgun hunter can account for quite a few woodpigeon with a little enjoyable roost shooting.

Decoying pigeon can provide good sport under the correct circumstances, although improperly used, decoys are useless. Full-bodied decoys can be lofted into trees to entice woodpigeon during roost shooting, and they may be lofted either with commercially available poles or with a simple piece of fishing line with a lead attached. The lead is either thrown or fired by a catapult over a branch, the 'deeks' tied on and drawn upwards. Decoying woodpigeon on the ground is another oft-quoted method of shooting woodpigeon, but it must be said that this is largely the province of the shotgunner for two reasons. Firstly, the birds will not stay on the ground very long as they soon realise that they have been 'had' and these are not real woodpigeon around them, and secondly because it is far harder to kill a woodpigeon on the ground that it is to kill one in a tree, due to the angle of the shot. However, a likely crop at the right time of year can attract woodpigeon, and if the farmer is stupid enough to let an airgunner put out deeks to attract even more birds onto his crop then there's no harm in having a go at this overrated sport.

When a pigeon comes in to decoys it will often land firstly in a tree if there is one near the decoys, and this behavioural trait may be used to good effect if the airgunner builds a bale hide within easy range of such a tree. If the tree is in a hedge then the hide may be easily built, though on no account cut wood to dress it - use grasses, ferns and fallen branches.

One way to account for woodpigeon during long, hot late summer days when the birds are feeding on corn is to hide near any handy drinking area such as pools or water-filled troughs. Corn makes birds very thirsty, and they will regularly flock to any water source. Decoys may be placed in the surrounding area, and as always they serve mainly to convince the pigeon that the area concerned is safe to land in. It pays to have at least one decoy in a position from which birds can spot it from a long way off, to draw in more birds than would otherwise visit the area.

To summarise, roost shooting gives the best opportunity for the airgunner to kill woodpigeon, although other methods can provide a degree of success at those times of day when roost shooting is unproductive.

Collared Dove

The numbers of this small bird have increased greatly in recent years and now they can be regarded as pests in certain circumstances. Not possessing the woodpigeon's instinctive fear of man, the collared dove will venture right into farmyards to raid grain stores, and they are often to be found almost resident around all kinds of rural dwellings.

Because the collared dove is such a petite creature, it is easily killed by almost any airgun in either calibre, and being easy to approach, it is very easily accounted for. Whenever the numbers of these birds warrant control, the airgun user has simply to put himself in the vicinity of the bird's food sources and wait for them to come to him!.

Building a Hide

Since a hide is so useful for shooting pigeons and crows, some notes on its construction are in order. First and foremost, when building a hide please do so with sympathy for the surrounding countryside: the result of many hide building exploits is a mess of plastic sheeting, garden canes and other rubbish, which not only constitutes an eyesore but which also presents a danger to farm stock - in particular cattle. If you take hide building materials into the countryside then please take them back with you afterwards.

In the summer hide building materials abound in the form of ferns, grasses, and so on, which may be woven into a framework made from dead branches. In the wintertime great reliance must be placed upon dried grasses, and a man-made net of some sort will often be a necessity. The beauty of utilising natural foliage is that it will match the colours of the season and be doubly effective.

You can buy many excellent hide nets - some complete with poles to allow the building of a tent-like structure, and it is not too difficult to dye material yourself, get hold of a few straight sticks (ash is superb) and build your own transportable hide. Always try and match colouring to the background, so that it looks 'natural'.

The Crows

Easily discernible by virtue of its all-black plumage, the carrion crow, unlike the rook, rarely flocks up with others of its kind, but is usually encountered solo or in a pair. The carrion crow is an opportunist feeder, and its love of another bird's eggs and fledglings has long placed it on the gamekeeper's 'hit' list. The carrion is a very wary bird and is often difficult to account for - the author has pursued individuals of the type in the past, and has often only taken them after a fair amount of scheming and with quite elaborate arrangements.

Unless a particular bird or pair is sought, the airgun shooter stands a chance of a carrion every time he sets out into the woods to shoot other crows, pigeons or squirrels. The same rules of concealment required for woodpigeon shooting apply when the carrion is quarry, and the bird will often warn of its approach with its harsh 'Wrraak' cry - once heard, never forgotten! Sometimes the bird will glide in on noiseless wings and alight close to the airgunner - scaring him half to death when it does decide to let out its noisy call!

There are attractants available to draw carrion, of which the best - a tethered ferret - is illegal, since the use of **any** living creature as bait for anything other than fish is highly illegal! A dead squirrel at the edge of a wood (where it is highly visible to a bird with such good eyesight as the crow) will draw in the bird, as will a dead woodpigeon or just about any other creature you care to think of. If you can set up a dead crow apparently feasting on the other cadaver then the picture becomes almost irresistible to the carrion. Alternatively, you can purchase decoys made specifically for the purpose: there is none better than the little owl decoy, the sight of which drives all crows to distraction. Always set up cadavers, eggs (these work well) or decoys in an area where they are most easily seen, and in a field at the edge of a wood near to a good, big oak for the crow to land in is ideal.

The method described for drawing crows will work equally well with magpies, jays and even rooks - although whether or not the latter bird really deserves the same sort of pest status as his near-relatives is debatable. The jay and magpie are birds of the woodlands and hedgerows, although both are generally spreading their habitats to include gardens, parks and even the open fields in some areas. The population levels of both birds are rising quite rapidly, posing a growing menace to song birds. Both the jay and the magpie are quire easy to actually kill with an airgun pellet properly placed: the magpie being a far smaller bird than most people believe and the jay paradoxically turning out on close inspection to be larger than anticipated. The main difference between the two is in their behaviour - the jay is the most shy and wary of the crows whilst the magpie is more of a 'head banger' - the avian equivalent of a football hooligan - often seen in groups of six or more and quick to join other magpies mobbing, perhaps, your little owl decoy. The only time the author has witnessed the jay abandoning its usual wariness is when it has been mobbing a grey squirrel - for which the jay appears to have a pathological hatred. If you are determined, therefore, to shoot jays then you will be well advised to set up a dead squirrel as an attractant.

Generally, it's best to set up an attractant, concentrate on camouflage for yourself, and settle down for sport with grey squirrels, magpies, jays and carrion crows - for all will respond to the same stimulants.

The other resident crows - the rook and the jackdaw - are not really pests in the way that other crows are - although rooks can make themselves a nuisance at times when large flocks raid crops. In such circumstances, most landowners will prefer to set up a bird scarer to letting a lone airgun user take the odd bird - although there will be exceptions and you can still find some farmers who will even pay for you to shoot rooks.

Chapter 9

Squirrel and Rat - Introduction

Both the grey squirrel and the brown rat are quite small animals and both live life at a fairly hectic pace, constantly on the move. The brains of these creatures are very small, and since these are highly mobile targets the hunter who uses a .22" to take them with body shots will fare far better than the .177" man who waits patiently for that elusive and difficult head shot.

The effect of a .22" pellet hitting such a small creature is not unlike that of a whole tin of .22" pellets hitting us full in the chest at about 400 miles per hour - devastating. Unlike the near-miss .177" head shot which strikes just outside a vital area the near-miss heart/lung shot with a .22" gives such striking force that a kill is most likely - and any rat or squirrel which survives such a shot will be incapacitated and quickly despatched.

One further point which while is does not go against the .177" for this shooting certainly facilitates the effective use of the .22" is that the ranges encountered are unlikely to be very long, so that there is not too great a need for the flat-shooting capabilities of the .177" which are so desirable in rabbit shooting. For squirrel and rat shooting, the .22" reigns supreme.

When looking for an air rifle for rat or squirrel shooting, one consideration arises which is not an important criteria in the selection of a rifle for rabbit or avian shooting, and that is the ease and speed with which the rifle may be re-loaded after a shot has been taken: since both the rat and the squirrel are capable of inflicting a painful (and in the case of the rat - a fatal - as we shall see later) bite, it is advisable to always make a practice of putting a shot through the creature's head at point blank range in order to be really sure it is dead. This is not 'cowardice' (as some anti-bloodsports people would portray the action): it is merely common sense.

The need for a quick second shot reduces the desirability of the in many ways excellent pump-up air rifle. This type of rifle offers recoilless operation, and is capable of great accuracy in the right hands, although to date there is no example of this type of rifle on the market which offers the shooter those hallmarks of the spring air rifle which we not so long ago dreamed of and which we now take for granted - good double pull triggers, good sights and robust construction. The spring air rifle has come a long way while the pump-up has remained a poorly-made alternative.

Reservoir rifles are ideal for rat and squirrel shooting, provided that the act of feeding a pellet is not too fiddly. All pre-compressed weapons are more efficient in .22" than in .177" (and even more so in .25"), and they do offer the person who holds a Firearms Certificate the opportunity to have a 30+ foot pound weapon, although the need for such high power in this type of shooting is questionable. Certainly a .22" pellet possessing even 18 foot pounds makes a fearful mess of either a rat or a squirrel, and one possessing 10 foot pounds or even much less down-range will do the business and give clean kills.

By and large the spring air rifle remains the better choice for this type of shooting. Most .22" spring air rifles will possess the necessary power to kill both rats and squirrels, most are robust enough to give many years of service to the most heavy-handed of users, and most offer reliability and ease of servicing and repair. Let's look in greater detail at the requirements of an air rifle for rat and squirrel shooting. Firstly, there is a great need for accuracy, as even the heart/lung areas of the rat and squirrel offer quite small targets - not much larger, in fact, than the head target of the much larger rabbit. This accuracy is dependent on several factors. Firstly, the rifle must have a usable trigger, by which I mean a trigger which is predictable in its release point. It matters not whether the trigger is single or double pull - in fact the humble single pull is often better suited to the kind of positive shooting needed for this sport. Often the rat or squirrel will stop moving for a very short interval, and the hunter has to bring his gun to bear and get the shot off as quickly as possible. There is still so much market resistance to the single pull trigger that we are unlikely to see any more air rifles being launched with this type of trigger. This is a strange trend, as the majority of sporting centrefire and rimfire rifles have single stage triggers the world over.

Accuracy is to some extent dependent on rifle/user compatibility. You are unlikely to shoot well with a rifle with which you feel uncomfortable, and if you have a rifle with which you have doubts then you'll never shoot well - psychology rears its ugly head. Physically, the rifle should be light enough for you to be able to hold it steadily, yet heavy enough for the same effect! Not only the overall but also the distribution of weight is important. Too muzzle heavy a rifle will be too slow to bring to bear quickly enough for some shots.

The only way to ascertain whether a particular rifle is compatible with your build is to actually shoot it for a protracted period, and the best place to do this is at a field target club, where you should find people willing to let you shoot their own rifles; few shops will tolerate a customer who spends a long time on their indoor ranges trying out guns.

What else to look for? The beauty of squirrel and rat shooting is that almost any spring air rifle is up to the mark, so that any rifle delivering in excess of ten foot pounds with average accuracy will be capable of giving clean kills at reasonable range. The choice is your's.

Sights

Open sights are perfectly adequate for squirrel and rat shooting in good light. Certainly for the grey squirrel open sights will suffice as the vast majority of shots at this creature will be taken in good light. With the rat many opportunities of shots will come in poor or artificial light, making the telescopic sight desirable.

Open sights offer the advantage of being very quick to bring to bear as the user has the widest possible field of view for tracking moving quarry, and if the rifle fits properly then as it is mounted the open sights will be pointing where the user is looking - at the quarry. As always, the foresight should be quite slim and the rearsight 'V' large to permit the best possible view of the quarry. Naturally, the sights should be adjustable, although once set, leave them and aim off to allow for distance and windage rather than constantly adjust the sights.

Most telescopic sights are a boon to the rat and squirrel

hunter - a few are a bane. Many novices have power on the brain - they want the most powerful rifle and obviously the most powerful scope - yet trying to locate a rat or squirrel through a powerful scope is often time-consuming, and by the time the scope is brought to bear on the relevant area, the beast has moved on! Small, highly mobile targets have to be tracked through a scope until they stop moving and permit a shot, and for this the lower the scope's power the better. Certainly a 4X scope offers the maximum degree of necessary magnification, and three or even two times magnification scopes are preferable.

One of the best squirrel and rat scopes I have used was my own version of Webley's 'Teleskan' system; a 1.5x15mm pistol scope mounted on the breech block of the rifle. This little scope allowed almost shotgun-style shooting - pure 'point and pull' - natural, and very fast.

Unfortunately, low power scopes usually have small objective lenses, and many novices like the macho appeal of a big scope! If you choose a 4x scope for general shooting, then you will not need one with an objective lens larger than 40mm, and these are probably the best practical choices for the majority of pockets. Low magnification scopes allow the viewer to look through the lens system from many angles and so permit parallax error to creep in, so making parallax correction very valuable.

Many people are now switching to so-called 'red dot' scopes which have an illuminated point in the centre of the reticule, and such scopes are quite useful; in normal light they may be used as ordinary scopes and in low light the red dot offers a very quick sighting system.

Pellets

Dispensing with .177" as quickly as possible, use the softest flat-head pellet you can find to take advantage of the lessened penetration such pellets offer against other designs. This will not match the capacity of the .22" for reliable body shot kills but it will make the most of the energy the pellet possesses.

Penetration, so often vaunted by pellet salesmen as a desirable characteristic of their wares, is the least needed trait for a rat or squirrel pellet. This is because a high penetration pellet will too easily pass straight through the body of the quarry, wasting valuable killing power. A heavy, low penetration pellet will not only reliably kill a rat or squirrel - it will bowl the rat over or knock the squirrel from the tree.

The absolute ideal rat and squirrel pellet in my experience is the flat-head 14-15 grain. A round-head pellet will kill squirrels and rats quite well enough, and whether you select a flat or round headed pellet should depend on which individual pellet gives the best accuracy through your rifle. The only way to ascertain which pellet this will be is to try several from the bench and compare groupings. Remember that even two identical rifles can favour different pellets, so don't blindly accept a recommendation from a friend who has the same kind of air rifle as your's. Don't accept any pellet which gives 'fliers' - you'll get few enough chances at rats and squirrels without wasting them due to poor pellets.

Chapter 10

The Brown Rat

All creation seems to at least dislike and at most hate and fear this small rodent which originated in Russia and which came to the U.K. in the early 18th Century. The brown rat was preceded by the smaller plague-carrying black rat, and whilst it won't spread the Black Death and wipe out another 25 million Europeans, as did its relative, the brown rat has a nice line in fatal disease called leprospiral jaundice - often called *Weil's Disease.*

Weil's Disease is a jaundice which will almost certainly kill you if you get it. Since the disease can be transmitted through the rat's saliva and urine coming into contact with a scratch or cut it's a fairly good idea never to touch a dead rat or to wrestle with a live one.

Weil's disease is horrible. I know it is horrible because my lurcher dog, Kai caught it after being bitten on the tongue by a rat in my garden, and his weight dropped from fifty-six pounds to just under thirty-six in the fortnight during which he hovered between life and death. Passing very large amounts of blood, Kai grew so weak that he had to be carried everywhere and supported while be urinated and defecated yet more blood. Massive antibiotic doses saved his life but left him very weak for the best part of a year - and this was an inoculated animal - imagine the effect of the disease on an unprotected human. Enough of this warning not to touch rats. If you don't heed it I won't attend your funeral!

Brown rats generally live wherever man allows them shelter and food, and although in the summer rats will often spread out along country hedgerows and live in the great outdoors the first frosts of winter usually find them snug near human habitation. As rats will eat almost anything, they can eek out a living from the most unpromising environments, and if all goes well for the rats, one breeding pair could produce twenty or more breeding pairs within a year!

A healthy doe rat can produce fifty young a year in five litters. and every rat will be capable of breeding within three months. Hence infestations of rats build up quite quickly.

2 photographs of dog
This is what Weil's disease did to a very healthy dog - it would probably kill you, so never touch a brown rat.

Unfortunately, few people would recognise an infestation on their property unless their dog suddenly died of Weil's Disease as the rat is by nature nocturnal and very shy of man. Those who can recognise the greasy slick of a rat-run will soon realise that their property has been invaded, but generally, rats live with little intervention from man simply because he does not know that he is playing host to these undesirable rodents.

Before looking at rat hunting with the air rifle, let's clarify one point - no-one controls rats with an air rifle. The best you can hope for is to account for the one wily old rat who is too smart to get caught in a trap or to take poison. Trapping, gassing and poisoning are the ways to control rats - air rifle hunting is a useful supplement, but not more.

The air rifle is useful for supplementing a poisoning regime as many poisons disorientate the rats and have a 'build-up' effect, which takes time, so that poisoned rats often appear in daylight against their natural tendency to nocturnal activity.

Apart from poisoned rats or chance encounters with displaced rats along hedgerows, most opportunities of rat shooting come at night under artificial lighting conditions. Within those farm outbuildings with electric lighting the best bet is to bait the rats for a few nights, leaving the lights on so that they become used to the new conditions in the building, and then to spend a night shooting in the building. If an infested building has no lights, then a hurricane lamp placed on an up-turned flowerpot set is a bowl of water (to prevent a fire if the lamp is knocked over) will suffice.

It is possible to use a normal lamping outfit for shooting rats indoors, although the natural reaction of a rat suddenly illuminated thus is to bolt for home, and success will be minimal. Rat lamping has been extolled at great length in the past by some people - let's hope this relegates the subject to its deserved position alongside moonlight rabbit shooting and other rubbish.

The one glaring exception to the nocturnal rat shooting norm is tip shooting. Rubbish tips are noisy places during the daytime with lorries coming and going and bulldozers busy levelling, and in my experience rats modify their behaviour on busy tips to be more active during the day than at night.

Tips are terrific places to hunt rat during the day. Bait them with some meat which has gone off and be prepared for some very fast shooting as the rats won't hang around for long. Baiting helps not only in bringing forth rats but also in getting them at whatever range you desire, so that the curved trajectory of the favoured .22" is less important. One point to note before setting out for your nearest tip: you must obtain the permission of the landowner to shoot anywhere - even for shooting rats on a tip.

Farm tips don't encourage daytime rat activity, though smelly enough bait will draw the rodents from their burrows at all times of day. Poultry draws rats, and the more poultry the more rats. Traps and poisons cannot actually be used in the poultry runs and sheds, of course (excepting battery houses), and so the air rifle is one of the best choices for rat control in such instances. Due to the problems of illuminating outdoor poultry runs and the uselessness of lamping, its best to try and bait the rats out into the open during daylight hours, and it might take a couple of days of baiting to get the rats confident enough for the eventual hunt.

Chapter 11

The Grey Squirrel

The grey squirrel, like all British mammal pest species, is an invader - this time from North America. The native squirrel, the small red squirrel, is nowadays protected and must not be shot. It is confined to a few areas mainly in Scotland, and so most people won't ever have to try and tell the difference between the grey and the red - for the record the smaller red squirrel has prominent ear tufts - they grey doesn't. In the winter, the grey does assume a russet colour along its back and so may at first sight appear to be a red - the ear tufts are the deciding factor.

The grey squirrel is largely arboreal (tree dwelling), although it does spend a lot of time on the ground, especially when food gathering. The grey squirrel also buries a lot of food during the autumn for consumption during the winter, although it does not find its caches too often and thus helps spread oak, beech and other trees. This one good deed is countered by the squirrel's more anti-social behaviour. Firstly, it often strips the bark from trees, especially during the snows of winter, to get at the soft inner sapwood which it eats. Secondly, it will readily take the eggs and fledglings of birds to supplement its diet. Thirdly, it is a great spoiler of fruit, and can wreak havoc in s small orchard.

The main habitat of the grey squirrel is woodland, although they have now spread into suburban gardens and parkland as well. Most control will be needed in woodland, however, as parks and gardeners often make welcome this engaging mammal - although the gardener's welcome is usually short-lived after the squirrels get into the fruit trees!

Grey squirrels breed twice a year, once in January and once in June. The young are born in a drey, which is generally a football-sized bundle of twigs. Summer dreys are much more flimsy than strong winter dreys and can be built further from the trunk than a winter drey. The average litter size is three, and the young are able to look after themselves at about ten week's age. Unfortunately, all this coincides with the end of the pheasant shooting season - a time when the gamekeeper turns his attention to controlling squirrels, with the result that young squirrels unable to fend for themselves can be left in the drey when the female is shot. I like to thin squirrels during late autumn/early winter, well before the breeding season, and generally the squirrels become shootable as they establish their territories and battle with neighbours and intruders. A noisy border dispute will attract your attention from quite a distance, and at such times the squirrels are generally so taken up with their skirmish that it is quite easy to approach.

Human considerations for young nest-bound squirrels apart, it is not difficult to hunt this rodent from autumn to spring - or when the leaf is off the tree and the squirrels are easily visible. When foliage thickens, treed squirrels can take some spotting, and even though they are highly active during such times a walk through woodland will let you hear far more squirrels than you see. Far better to look for a food source of some kind, and wait in ambush. Squirrels tend to establish regular runs a to and from feeding areas, and the alert hunter will usually be rewarded if he does his homework properly and locates a good run.

Hedgerows are excellent places for ambushing squirrels if they run from the holding woods, especially if they lead to a good food source such as farm corn stores or any buildings holding stock. Squirrels will return again and again to a proven food source, and one good vantage point near a well-used hedge can provide many shots. Look for a point along the hedge where the squirrel tends to stop, such as before it has to make a jump to the next tree (the squirrel won't stop at such places if it has seen you, however). If you don't know where a squirrel might stop, then either get hold of a squirrel call or learn to make a noise resembling a rabbit squeal - the noise will make the squirrel stop.

Woodland shooting for grey squirrels is so highly regarded as a sport in America that the animal is awarded a closed season in order to protect it during its breeding season! Americans tend to hunt squirrels with shotguns or rimfire rifles, but here in the U.K. the safer bet is the air rifle: we simply don't have the vast empty tracts of land which in America permit the safe use of the rimfire, which can kill humans at a range of 1 mile in some circumstances.

Woodland squirrel hunting can be approached in several ways. The hunter can simply go for a pleasant walk through the woods and take his shots as and where they appear. He can sit down almost anywhere and the squirrels will sooner or later come to him. He can search out feeding areas, routes or the areas where the squirrels live and sit down to await action. Whichever method is chosen, if there are squirrels around then shots will be on.

Squirrels live in trees - everyone knows that, but many novices fail to realise that the grey squirrel spends a lot of its time at ground level whether foraging or travelling, and the novice is inclined to spend his days a'wood with his neck craned to espy the treetops when the squirrels at ground level easily see his stumbling approach and scatter before he sees them. Two pairs of eyes are best for squirrel shooting, and the second pair can be either a companion or a dog - most dogs take to squirrel hunting with a verve, and since their eyesight is often better than our own, they can be a great benefit. Squirrel hunting, incidentally, is about the only time I really recommend the airgun hunter to take a dog with him - as dogs don't go for all the stalking and sitting necessary for other airgun quarries. My own dog never ceases to amaze people I take hunting with his usefulness for squirrel - he won't just spot them for me - he retrieves and sometimes catches them as well - that's keen!

When you reach woodland, it's worth remembering that squirrels are most active near the fringes of the wood, and circumnavigating it before actually going in can give quite a few shots. Any junction between wood and open ground attracts squirrels, and any rides through the woods will see plenty of squirrel activity.

Actually inside a wood, pheasant feeding hoppers and release pens attract squirrels (plus jays and magpies) regularly, and provide many shots for the airgun hunter who is friendly enough with a keeper to be trusted in the coverts! There is a very strong case for shoot owners to let an airgun hunter into their woods. Apart from culling squirrels, rats and corvids, the airgunner is a presence which can help keep professional poachers at bay.

Unfortunately, most game shoot owners and gamekeepers are paranoid about their birds and won't trust anyone but a fully paid-up gun on their land. If you happen to be a keeper or a shoot owner and have bought this booklet to see whether the airgun can be used to cull squirrels and rats near the coverts then the answer is a resounding YES! The airgun is the number one tool for such work, and it is so quiet that you won't scatter your birds when you shoot. Please spare a thought, though, for the airgun hunter who is often desperate for shooting, and who can be so useful a guest on your land.

There's more to squirrel shooting than merely seeing the beast and pulling the trigger. For one thing the squirrel quite rightly has a desire not to be shot, and will do anything to hide from the hunter or put distance between himself and the hunter. If the squirrel sees you before you see him, then he will either put a tree trunk or a branch between him and you, and cling onto the far side.

Fortunately for us but unfortunately for the squirrel, this action usually leaves his tail visible, and when sunlight strikes the tail it becomes highly visible. As a squirrel stands more chance of seeing you first than you do of seeing him, you need to develop an ability to spot that bit of tail if you're to achieve consistent success as a hunter of squirrels. A treed squirrel which is hiding from you will move around as you move to try and gain a better view of it, and it will normally keep enough branch in between it and you to prevent a shot. In such circumstances you rely on your dog or your shooting companion to move to the other side of the tree so that the squirrel in turn moves towards you. If you're on your own try taking off your jacket, hanging it on a bush, and moving slowly to where you can get a shot while the squirrel hopefully concentrates on hiding from your jacket.

There are times when the squirrel is less than alert and may be stalked in addition to the territorial disputes already mentioned. Whenever a squirrel is foraging for food or burying excess food then its wariness reduces greatly, so consumed is it with the job in hand. From time to time it will look around and sniff the air for danger signs, but you should be able to overcome this by stopping your movement when the squirrel's head comes up.

Shotgunners sometimes use long poles to poke dreys and get the squirrels out into a shootable position, which is, of course, a useless practice for the airgun hunter, who requires a stationary target. Anyway, I practice, as do the Americans, conservation where the grey squirrel is concerned, for its pursuit is such good sport that I want to save some for the future.

Like the rabbit and the brown rat, the grey squirrel is an invader who has been with us for so long that his demise would be a sad day. With no natural predators worthy of mention, a bi-annual breeding season and a potential lifespan of up to ten or so years, the grey squirrel will probably be with us for many years to come. As for the brown rat - he'll probably be around forever.

Chapter 12

Other Species

Several species not already covered - both on and off the MAFF vermin list - may be reliably killed by a skilled air rifle shooter, though whether or not to tackle them is a matter for personal deliberation.

The brown and the blue hare can be killed with the air rifle provided that the shot strikes the cranium. The author has in the past taken both types of hare with the air rifle but cannot recommend the practice. Firstly, the brown hare cannot now be considered vermin (though there are reportedly infestations of blue hare in parts of Scotland). Secondly, unless the shot is perfect then this powerful animal will escape wounded - something to be avoided. The author used to always take his lurcher along when hunting hare, as the dog could have caught any wounded beast - though the need never arose.

The mink is on the MAFF list and it is possible to kill mink with a well-placed pellet to the cranium. If you have occasion to visit a river or stream where you have shooting permission and where mink are known to exist then it is worth taking an air rifle along with you. Unless you have shooting permission on both banks, however, safe and lawful shooting will be very difficult and you should avoid shooting onto or over water for rear of ricochets, anyway.

Other smaller members of the mink's mustelid family - the weasel and the stoat - could be killed with the air rifle, though these creatures appear to be becoming rarer as time goes on and whilst they might have in the past been killed because they posed threats to the well-being of game birds I cannot endorse the activity nowadays. On game estates it was once common practice to cull raptors - hawks. These birds now enjoy the fullest protection of the law and may not be killed under any circumstances.

The air rifle is physically capable of killing game birds, though I can think of no landowner who either considers them vermin or who will entertain the notion of them being shot with an air rifle. Not recommended, though if you own land and want a pheasant for the table and possess a licence then there is no reason why you should not do as you please.

Never entertain the idea of pursuing the fox with an air rifle. The rifle itself is far too underpowered to kill foxes and all you could achieve is a wounding. If you have a fox problem then the rimfire rifle is the best tool for the job.

Chapter 13

Other Types of Gun

Whilst there are many circumstances in which the air rifle can be the most appropriate gun for vermin control, there are instances where other types of gun can be preferable.

Shotguns

The shotgun is defined in law as a smooth bored weapon with a barrel length not less than 24". By law, the cartridge used must contain at least five pellets - otherwise the shotgun becomes a Section One Firearm.

Unlike the air rifle and rifle, the shotgun has no sights as such, and usually possesses only a small bead at the muzzle. This is because the shotgun is not 'sighted' in the same way as a rifle, but is instead pointed at the target. The shotgun fires anything from five to over two hundred small pellets, which spread the further they travel from the muzzle to give a pattern which might typically be 40" in diameter at forty yards. Because of this spread, the shotgun may be used - in fact is best suited for use against - moving targets such as birds on the wing or a running rabbit. There simply is no time for careful aiming; the gun is mounted, brought onto target and discharged very quickly.

The shotgun possess characteristics quite unlike those of the air rifle, which influence the circumstances in which either might prove the better choice for vermin control. The shotgun possesses massive energy in comparison with the air rifle, and so may be used against larger quarry such as the hare and fox, whereas they very low power of the air rifle obviously means that it may be used in many instances in which the sheer destructive power of the shotgun renders it too dangerous. The quiet operation of the air rifle, which can permit many shots to be taken in a small area, is in marked contrast to the loud report of the shotgun. If crop protection is the reason for shooting then the shotgun is obviously the better choice.

Shotguns can be single or double barrel, or of the

pump/auto kind. Pump and auto-loading shotguns are to be restricted to two cartridge at the time of writing, though generally shotguns may be held and used by anyone possessing a shotgun certificate, which is not too difficult to obtain. Ask your local gunsmith for details.

Shotguns come in a wide range of calibres. Almost useless are the .22" and 9mm 'Garden Guns' - capable of only short range work against the smallest of vermin. The smallest practical calibre is the .410", which fires around half an ounce of shot and which is capable of killing all air rifle quarries, albeit at closer ranges than the airgun's maximum. Apart from the .410", most other shotguns bore sizes are designated by the number of lead balls of the bore diameter which weigh one pound. Hence twenty lead balls weighing between them a pound would fit a 20 bore, and so on.

The 28 bore shotgun is a small step up from the .410" and has a similarly restricted useful range. Further, the cartridges will often be difficult to obtain, as the calibre is not at all common. One of the most useful of calibres is the 20 bore, firing up to one ounce of shot. This gun has reasonable recoil and quite useful performance. The 16 bore is a calibre much favoured on the continent, but quite rare here with consequent higher ammunition prices and lessened availability. The 'standard' U.K. shotgun is the 12 bore, which usually fires between one ounce and one and three-sixteenths of an ounce of shot. Larger calibres exist in the main for wildfowling.

Rifles - Introduction

There are essentially two types of rifle: rimfire and centrefire. The rimfire is a far lower power weapon than the centrefire, and when you consider that even the *little* rimfire can kill a human being at a range of one mile then the awesome power of the centrefire which rules out its use against anything other than deer on vast areas of forestry becomes apparent.

The rimfire is the next step up from the air rifle, although its power is usually anything between seven and thirteen times that of the air rifle. The rimfire is useful against hare and fox, otherwise it is best considered as the weapon which takes over when the useful range of the air rifle runs out. Some will argue, but the rimfire is a 100 yard weapon with typical velocities of between 800 and 1,400 feet per second and bullet weights up to 40grains.

In order to possess a rimfire rifle you require a Firearms Certificate and a lot of unpopulated land!

Chapter 14

The Rimfire Rifle

The rimfire rifle offers a step up in performance from the air rifle and the two weapons are in fact complementary on each other. For this reason, it was felt appropriate to cover the subject of buying a rimfire rifle.

If you use a rimfire then I would advise that it is kept solely for use against ground game: shots at tree-borne quarry constitute a great danger to the public as the bullet can easily pass right through the body of the quarry and continue on to kill a person up to a mile away.

In order to own a cartridge rifle it is necessary to obtain a Firearms Certificate from the police Firearms Department. Application forms are available at all main police stations, and certain conditions will be imposed on successful applicants. Firstly, the rifle and ammunition will have to be stored in a locked metal gunsafe when not in use, and such safes are readily available at gunshops. Secondly, the certificate will state where the firearm or firearms to which it relates may be used. If the applicant can gain permission over enough areas of land then he may be granted an 'open' certificate which allows him to use the rifle over any land 'deemed fit by the Chief Constable for the area'. The police visit any area of land over which a cartridge rifle is to be used, and either pass or fail it for such usage.

Before applying for a firearms certificate, therefore, it is best to obtain a security cabinet for the safe storage of the rifle and ammunition, and also to sort out some land over which the rifle may be used. If the land is not already passed for rimfire use then make sure that it is safe before you apply.

A Firearms Certificate will state exactly how many of which types of weapon may be kept by the holder, and it will also stipulate the quantities of ammunition which may be both purchased at any one time and held at any time. The serial numbers of any weapons will have to be entered on the certificate, and, as we have already established, it limits the areas in which the rifle may be shot.

Once you have been granted a Firearms Certificate for a

rimfire rifle then you will want to acquire a rifle as quickly as possible. The best piece of advice is to tread warily, as if you buy wrongly then there's a ridiculous amount of red tape attached to changing your mind! In such instances, you must surrender the weapon and send your certificate off to be altered - a process which can leave you without a rifle for weeks. When the certificate is returned then you can get another rifle. With so much fuss attached to the simple act of a one-for-one swap, the reasons for getting the right rifle the first time are obvious. The question is, which rifle?

There are many cheap second-hand rimfire rifles on offer - some can be good buys, others are a waste of money. Remembering the problems of changing your mind if you buy the wrong rifle for your needs it's best to very carefully consider whether you might be better to shun the old £5 rifle and blow a couple of hundred on a new one. Unless you know what to look for in a rimfire rifle I would recommend the latter.

Rifles which are available cheaply include old BSA Sportsmans, target 'converted' Martini actions, pump rifles intended for gallery use, and all sorts of odds and ends which have found their way onto the market over the years. Buying any is a risky business unless you can get the rifle onto a range and test the accuracy before parting with cash. Remember that an inaccurate rifle is as much use as a car without a steering wheel!

New or used, there are several types of rimfire rifle on the market. Most common is the bolt action, which can be either single shot or have a magazine to facilitate multi-shot capabilities. Almost as common is the semi-automatic rifle, which has an action which cycles when a shot is taken and feeds a fresh round into the chamber as well as cocking the action. There are many pump-action rifles on offer, although many of these were intended mainly for gallery shooting. Least common is the Martini action: most examples of this type of rifle are converted target rifles which may have been good in their day but which are often shot-out.

When choosing a rifle many novices plump for the auto - for the wrong reasons! People think that they can keep a rifle at their shoulder and pump many shots into, say, a fox, whereas with a bolt action this is impossible: true, but if you're any sort of shot you should not *have* to keep on pumping bullets into your quarry - one shot should do the 'business'. There is a problem with the auto rifle which is not apparent until you actually encounter it in the field, When you take a shot with a bolt action rifle the rifle is immediately safe - it has an empty round in the chamber. Take a shot with an auto, though, and you've got a tiger by the tail inasmuch as you still holding a loaded rifle - and trying to make that rifle safe in the dark when you have to collect your rabbit is too complicated an operation. I favour the bolt action for this single reason: until you have tried wrestling with an auto in the dark to make it safe then you cannot imagine how frustrating an exercise it can be.

In choosing a rimfire rifle for field use most economies are false economies. The *bargain* rifle usually turns out to be too inaccurate to ensure clean kills. The small bullets found in rimfire cartridges do not possess either the knock-down power nor the outright energy to kill with anything other than a head or heart/lung shot. As the quarries sought by the rimfire user will be no larger than a fox and often much smaller the need for accuracy becomes obvious.

A new rimfire rifle will probably cost in the region of £120-£200, although you can pay much more for a top-class rifle. Add to this the price of a good telescopic sight and mount, the Firearms Certificate and a security cabinet and you will be looking at a total cost in excess of £400. Many people will be forced to make economies somewhere along the line, perhaps doing without a scope or building their own security cabinet, though most people will begin their economy with the rifle itself.

When looking at a second-hand rimfire rifle there are a number of points to be considered. First, think about the overall appearance of the rifle. If it is covered in rust and the stock shows dents and scratches then the rifle has obviously has a hard life and seen little in the way of care - and the bore will generally be similarly rough. Next, look at the state of the bore. If the rifle is a bolt action simply remove the bolt - if it's an auto, pump or martini action then open the breech and place your finger nail at the end of the bore so that light is reflected inside - allowing you to look from the muzzle (if the barrel is too long for this then get someone to hold a piece of white paper in the breech for you).

You will probably have to clean the bore before you will be able to actually see metal. and this is an operation best done as follows. You will need a rod, a phosphor-bronze brush and a jag, and some Hoppe's No. 9 or a similar solvent. Firstly, run the phosphor-bronze brush through the barrel a couple of times to remove loose fouling and loosen the rest. Then dip the brush in the solvent and repeat the operation taking the greatest care not to get the solvent onto your skin or into your eyes. Leave the rifle muzzle-down for as long as practicable for the solvent to act (an hour at least, but preferably overnight). Run the jag fitted with a clean patch through the bore and marvel at the amount of gunge it fetches out! Repeat the operation once or twice more and you have a bore which may be examined.

When viewing the bore, look firstly (obviously!) for any pitting, and if the bore has even the slightest pitting evident then reject the rifle there and then - it is useless. Look also at the actual rifling. Old rifles are often *shot-out* - the rifling has simply worn away through too much usage. Rifling should be crisp in outline and fairly prominent in most barrels, although 'button rifling' (very shallow rifling) features in some rimfires and can confuse the issue. If you're at all in doubt regarding the state of the rifling then it is safest to reject the rifle - no matter how much of a bargain it may appear to be.

Now turn your attention to the trigger. There are just two types of trigger - good triggers and junk. A good trigger is any which lets the user know exactly at what point in the travel the rifle will discharge - a useless trigger has a degree of *creep* into the release point so that the user has to take a mighty pull on the blade which normally pulls the rifle off-target. Making doubly sure that the rifle is unloaded, try the trigger a couple of times. If it has just a single stage of travel then there should be absolutely minimal travel to the release point - if it is of the double pull type then it should release immediately after the *stop* in the travel is passed. For the record, I prefer the single pull trigger to the double pull.

Ensure that the trigger is safe by again cocking the unloaded rifle and giving it a few shakes and hitting the butt with the heel of your hand (ensuring that it is pointing in a safe direction). If the trigger releases then reject the rifle. An unsafe trigger may be corrected by a competent gunsmith, as indeed may be a bad trigger, but as the state of the trigger is indicative of the state of the rest of the rifle my advice is to reject it.

If you can, then remove the action from the woodwork and take a close look at the trigger and the metal which is normally sheathed by the stock, The state of the finish here is a good indicator of the overall quality of the rifle. The trigger, like the whole of the action, should be clean and free from dust, dirt and rust. There should be a minimum of lubrication apparent, as too much oil serves only to collect dust.

When the action is worked by hand it should be smooth and not *grind*. Look for wear marks on all bearing action components. Really, unless the rifle is actually new then you should shoot it before you commit yourself - which means in law visiting an MOD-approved range with the owner and firing to assess both the functioning of the action (doubly important in the case of an auto) and the accuracy. To buy without trying is one hell of a gamble.

If all seems well with the rifle then seek the importer and ascertain whether the model is current and whether spares are readily available. Imagine losing a box magazine for a rifle which is no longer made! It's best to avoid obsolete rifles unless you really can be sure of obtaining spares easily.

Buying a new rimfire rifle is by no means a straightforward business, as the sales of rimfires are rarely high enough to finance extensive advertising so that it is difficult to ascertain exactly what is available. Even when I worked in the gun trade and had access to the trade catalogues of main importers it took a fair amount of detective work to prepare a list of rifles and prices for comparison - if you're not in the gun trade then the position is far worse! Most people will be content simply to visit a gunshop and buy whatever happens to be in stock at the time, so depriving themselves of a wider choice.

Your local gunsmith should be able to get a list of available rifles together for you.

Ammunition

There are many different types of rimfire round, from the tiny B.B. cap to the 1600 feet per second Stinger and other *hot* rounds. Most rimfire rounds have uses, although some, it must be said, are useless.

The least useful .22" round is the Long Rifle Shot cartridge, which contains a few tiny *dust* shot. Effective against mice at ranges up to about six feet, the use of this ammunition through a rifled barrel will give only the periphery of a circle as centrifugal force affects the shot as it spins up the barrel! The round is virtually without use.

The smallest solid round is the B.B. cap, which contains a minimum of propellent and either a round ball or a conical bullet. The round ball tends to fall from many of the cartridges before they even see a rifle, and those which do find their way into the breech will bring many disappointments because their velocity varies so wildly. Consecutive shots with a .22" round ball gave velocities ranging from over 800 feet per second to 400 feet per second! The faster bullet will fly a full 4" higher than the slower one at a mere 20 yards range, giving hopeless inaccuracy. Like the shot cartridge, these are fairly useless.

The smallest *useful* rimfire round is the .22" short, and, as the name suggests, these have a short case. Developing around 45 foot pound's worth of muzzle energy, the .22" short is rather more powerful than the highest power air rifle, and emerges as a useful rabbit round by virtue of its quietness in use - especially for those who have to cull rabbits but who don't possess a silencer.

I have taken many foxes using the .22" short at close range by treating the quarry as a rabbit and hitting it in the cranium, or brain case. The main drawbacks of the short cartridge are the curved trajectory and the fact that they won't feed through all rifles.

One very useful round is the LR (long rifle) pistol, which usually gives a velocity of around 850-900 feet per second with a 29 grain bullet. This is a very quiet and accurate round which forms a stepping stone between the .22" short and sub-sonics.

Going up in power we find the .22" Subsonic, a round

which gives velocities under the 1,100 feet per second sound barrier. This round gives very quiet operation in silenced rifles, whereas supersonic bullets will still give a loud crack even when a silencer is used. With bullets ranging roughly between 29 and 40 grains in weight, the rounds usually give between 70 and 100 foot pounds.

In between the sub-sonic and high velocity rounds come a range of rounds giving roughly 1200 feet per second with 40 grain bullets. As these won't be quiet when used with a silencer, most sporting shooters tend to choose either the sub-sonic for silenced use, or the higher velocity rounds for foxing. The 'in-between' rounds, though, are often intended for target use and are of very high quality. Many more are intended mainly for the US market, where silencers are unobtainable and sub-sonic rounds are thus rather thin on the ground.

High velocity rimfire rounds tend to achieve between 1400 and 1600 feet per second with a 30 grain bullet for a muzzle energy of roughly 150-200 foot pounds, though their usefulness is related more to their flat-shooting properties than their energy, which is still puny when compared to centrefire rounds. At fifty yards, such rounds will drop around 3-4", and the better examples (40 grain 1400 fps rounds are better than 1600 fps 30 grain rounds at distance) will give reasonable accuracy at 100 yards when used in a good quality and compatible rifle.

They key to choosing an ammunition for your rimfire rests with the compatibility twixt rifle and cartridge to a large extent. Obviously, if you have to shoot at longer ranges then a higher velocity round is preferable, but if you cannot hit the small kill area at range then a cartridge is useless. The only way to find a compatible round is to test as many ammunition types as possible from a bench rest at paper targets, choosing that which gives the best accuracy.

Chapter 15

Nets and Ferrets

As Autumn slowly gives way to winter and temperatures by day drop and night drop, the rabbit's breeding rate slows down, and any does which are carrying young can, should the weather become harsh enough, re-absorb the young into their bodies (resorbing) to take advantage of the nutrients. To my knowledge, the rabbit is the only creature able to do this, and it is an important factor in the rabbit's survival, as in addition to ensuring that young are not born in conditions which will be too harsh and kill them, it makes the does very strong and well able to withstand the cruel winter weather.

As Winter draws nearer, the lush Summer undergrowth dies back, so allowing us a look at the many hedgerow warrens which have been concealed largely by nettles for the past six months, and the combination of no kittens and free access to the warrens signals the start of ferreting in the country calendar.

It is important that there should be no kittens in the burries when a ferret is entered because the ferret easily catches and kills them - generally drinking their blood and perhaps eating a little before curling up and going to sleep! this is called *lying up,* and we'll look at how to deal with it a little later in the chapter.

Many readers will already keep and work ferrets, but for the uninitiated, they are small members of the mustelid family, closely related to the stoat, weasel, otter, mink, polecat, pine marten and badger. The ferret is a domesticated polecat, and like the modern domesticated dog it bears little resemblance to its ancestor, being generally much smaller than a polecat, and often creamy coloured in contrast to the darker markings of the poley. Ferrets smell - handle one and you'll carry the aroma of ferret for some time afterwards - and to a rabbit which has a far more highly developed sense of smell than we do the ferret must positively reek! The polecat, stoat, weasel and ferret are great hunters of rabbits, and have been for aeons, and as soon as a rabbit underground gets a whiff of the old mustelid smell it starts to get very worried, and the sounds of a ferret actually in the warren will normally make the rabbits exit by a bolt hole (the origin of that expression, I would think) and use their speed to flee the predator above ground. The ferreter takes advantage of the rabbit's fear of the ferret by putting his ferret into a warren to get the rabbits to bolt.

Once they have bolted, there are several ways in which the rabbits may be caught. The simplest is to cover all holes with nets known as purse nets because they have a draw string which closes the net once a rabbit has hit it. Sometimes, the ground may be unsuitable for purse netting because of brambles, and then the ferreter might use a shotgun to shoot the bolting rabbits, he may course them with lurchers, or he may set a long net around the burry. A long net is, as the same suggests, a very long net which is set using a series of stakes perhaps 2'6" above the ground. If a rabbit runs into the baggy part of the net it becomes entangled, and may be taken out and dispatched.

Large, open burries in the middle of a field or on an undergrowth-free bank may be shot with an airgun to ferrets. Simply, the airgunner creeps over to the burry and enters his ferret and then quietly returns to his airgun twenty or so yards away. He then lies down and waits for the rabbits to start moving. Often, they will simply try to evade the ferret by popping up out of one hole and then diving down the next, pausing for a few seconds above ground in between - allowing sufficient time for a shot. The airgunner should be prone so that his silhouette does not immediately further frighten the rabbit. This is great sport, and allows rabbits to be taken from burries which are so open that they are normally unapproachable.

When the ferret manages to catch and kill a rabbit underground, it will often lie up, and in order to retrieve it the ferreter must either dig it out or try and move it with a liner. A liner is a large ferret which will drive the smaller jill (female) often used for hunting off its kill. As the liner is attached to a length of twine (the line), the ferreter can sometimes withdraw it if it's close to the edge of the burry, or more often he will have to dig the liner out, judging where to dig by the amount of line taken into the burrow by the liner.

Photograph of people
A B.A.S.C. vermin control course.

The electronic age has given us the ferret finder, which is a small transmitter affixed to a collar worn by the ferret, and a receiver which gives a stronger bleep the closer it is to the transmitter, so allowing the ferreter to find the position of his ferret. Good ferrets rarely lie up, and many people only work younger ferrets because they think they're less likely to catch and lie up than an older, wiser ferret.

If you decide to have a go at ferreting, the best time to buy a ferret (always a young one - no-one sells a good adult worker) is during the Autumn. This gives you plenty of time to get the ferret used to being handled before work starts later in the year. You'll need a hutch about the size of a rabbit hutch, in which you must supply clean drinking water. The ferret may be fed on any meat, or milk, bread and egg an an alternative. If you are any good with an airgun, though, you should have no trouble keeping your ferret in meat during the summer months!

Read 'Modern Ferreting' by my friend D. Brian Plummer. If you can't afford to buy the book then your library will get it for you for a small charge. This book will tell you everything you'll need to know in order to add new dimensions to your rabbit control, and if like me you wish to control rabbits during the winter when it's too cold to sit around for long with the air rifle then the ferret will provide the means.

Chapter 16

Other Rabbit Control Methods

The most controversial method of rabbit control is inarguably the most efficient; snaring. A snare is simply a wire noose which is set on a rabbit run in such a way that the rabbit's head passes through the noose: if the rabbit is travelling at speed, then the snare should break the neck.

Snaring has many drawbacks. Most importantly the snare should only ever be used by highly experienced persons as not only is there a considerable art involved but a poorly set snare will be a danger to all wildlife. Secondly, snares should never be set anywhere in domesticated animals might be encountered.

All in all, I cannot recommend the use of snares other than in exceptional circumstances - and then only by an experienced person. It is unfortunate that snares may be bought so cheaply, as this encourages inexperienced people to try them.

The second most effective rabbit control method is gassing, using a cyanide in powder form which is placed in the burrow and the burrow then sealed with a turf. Cyanide powders give off lethal gasses which can kill the person just as easily as a rabbit - never gas rabbits alone, as if you do pass out then your companion will literally save your life by dragging you away from the fumes. Don't use gassing methods unless you can find an experienced person to show you how.

You can trap rabbits using approved traps - not the now illegal gin trap. Like snares, traps are cheap to buy and inexperienced persons could severely injure themselves simply through not knowing how to set the trap. By law, all traps must be set under a cover which prevents other species from being caught; like snaring, trapping is a matter for the experienced.

The British Association for Shooting and Conservation run vermin control courses which cover both gassing and snaring. Contact their Education Officer for details.

Appendix

Useful Addresses

British Association for Shooting and Conservation (BASC),
Marford Mill, Rosset, Wrexham, Clwyd, LL12 OHL
Tel. 0244 570881

Organisation for all live quarry (non-target) shooters.
Membership gives third party shooting insurance and
quarterly magazine. Represents the interests of the shooter at
all levels. Many courses organised on vermin control,

British Field Sports Society (BFSS),
59 Kennington Road, London SE1.
Tel 01 928 4742

Organisation for all fieldsports, especially hound sports.

Airgunner. Editor - Paul Dobson,
2, The Courtyard, Denmark Street, Wokingham, Berkshire,
RG11 2LW
Tel 0734 771677

Monthly airgun magazine. Covers vermin control, field
target, plinking. Rifle and accessory reviews, articles on
collecting, airgun improvement etc.

Airgun. World Editor - John Fletcher,
10 Sheet Street, Windsor, Berks, SL4 1BG
Tel 0753 856061

Monthly airgun magazine. Covers vermin control, field
target, plinking. Rifle and accessory reviews, articles on all
aspects of airgun sport.

Shooting News. Editor - Clive Binmore,
Unit 21, Plymouth Road Industrial Estate, Tavistock, Devon,
PL19 9QN

Weekly newspaper, covering all fieldsports. Monthly airgun
special feature with hunting articles.

Shooting Times. Editor - Jonathan Young,
Address as Airgun World

Weekly magazine dealing with all aspects of field sports.

Sporting Gun. Editor - Robin Scott,
EMAP Pursuit Publications Ltd., Bretton Court, Bretton,
Peterborough, PE3 8DZ
Tel 0733 264666

Monthly shotgun magazine.

INDEX